Management for Professionals

The Springer series *Management for Professionals* comprises high-level business and management books for executives. The authors are experienced business professionals and renowned professors who combine scientific background, best practice, and entrepreneurial vision to provide powerful insights into how to achieve business excellence.

More information about this series at https://link.springer.com/bookseries/10101

Judith L. Walls · Andreas Wittmer
Editors

Sustainable Aviation

A Management Perspective

Editors
Judith L. Walls
Institute for Economy and the Environment
University of St. Gallen
St. Gallen, Switzerland

Andreas Wittmer
Center for Aviation Competence
University of St. Gallen
St. Gallen, Switzerland

ISSN 2192-8096　　　　　　　ISSN 2192-810X　(electronic)
Management for Professionals
ISBN 978-3-030-90894-2　　　ISBN 978-3-030-90895-9　(eBook)
https://doi.org/10.1007/978-3-030-90895-9

© The Editor(s) (if applicable) and The Author(s), under exclusive license to Springer Nature Switzerland AG 2022
This work is subject to copyright. All rights are solely and exclusively licensed by the Publisher, whether the whole or part of the material is concerned, specifically the rights of translation, reprinting, reuse of illustrations, recitation, broadcasting, reproduction on microfilms or in any other physical way, and transmission or information storage and retrieval, electronic adaptation, computer software, or by similar or dissimilar methodology now known or hereafter developed.
The use of general descriptive names, registered names, trademarks, service marks, etc. in this publication does not imply, even in the absence of a specific statement, that such names are exempt from the relevant protective laws and regulations and therefore free for general use.
The publisher, the authors, and the editors are safe to assume that the advice and information in this book are believed to be true and accurate at the date of publication. Neither the publisher nor the authors or the editors give a warranty, expressed or implied, with respect to the material contained herein or for any errors or omissions that may have been made. The publisher remains neutral with regard to jurisdictional claims in published maps and institutional affiliations.

This Springer imprint is published by the registered company Springer Nature Switzerland AG.
The registered company address is: Gewerbestrasse 11, 6330 Cham, Switzerland

Foreword

SITA, or Société Internationale de Télécommunications Aéronautiques, was founded more than 70 years ago and is the world's leading specialist in air transport communications and information technology. Serving more than 2500 customers in more than 200 countries and territories, SITA is one of the most diverse international companies. SITA has always been on the forefront of technological breakthroughs and is continuously following industry trends to ensure that it is aligned with its customers' needs. As such, SITA has placed sustainability as a key priority for its business going forward. SITA will achieve carbon neutrality of its own operations before 2022 and become one of the first companies in the aviation industry to do so.

SITA for AIRCRAFT is a business unit of SITA providing services to airlines, aircraft and equipment manufacturers and air navigation service providers with the aim to make aircraft operations safe, cost-effective and environmentally sustainable. SITA for AIRCRAFT is actively contributing to aviation sustainability goals which was our main driver for this collaboration with the University of St. Gallen in Switzerland.

Aviation has been the foundation of global development and is essential for human connectivity. Aviation brings people and cultures together and it provides work to millions of people around the globe. Prior to the COVID-19 pandemic, the industry enjoyed unprecedent levels of growth, particularly in the last 20 years. Such growth and success certainly come with a responsibility to minimise and avoid negative impacts on the environment. The aviation industry has recognised that responsibility early by introducing the first ever industry-wide initiative to tackle CO2 emissions. However, the presumed high level of future growth, as well as capital intensity and technological complexity to decarbonise, is exposing aviation to the risk of becoming a key polluter in the transport segment in the next 30 years. Therefore, it is of utmost importance to find solutions that will fast-track industry's sustainability efforts. Only then can the industry continue to grow and further democratise the mobility of people and goods.

There are a growing number of aviation decarbonisation road maps that governments have put forward with the goal to eliminate carbon emissions by 2050. All of these plans include improvements in aircraft technology, operations and fuels. According to these road maps, operational improvements can contribute to up to 6% of overall aviation emission savings. If we consider the impact of other non-carbon emissions like nitrogen and clouds, caused by contrails, the benefits that operational improvements can bring are potentially much higher.

Therefore, besides insisting on longer term solutions, it is crucial to embrace technology that will provide benefits today. At SITA for AIRCRAFT, we are providing technology-based transformative solutions which enable safe and efficient flights. For example, we provide information to pilots on how to select more optimal flight levels and routings based on most actual weather information. This enables them to save fuel, reduce emissions and improve passenger comfort and safety.

As we transform our business, by integrating sustainability in its core, we fully understand the challenges and barriers that all businesses are facing. Consequently, we strive to support all activities that can ease the transition to more sustainable business models. We focus on the practical approach to sustainable aviation, looking for solutions that are open for all with easier transition, yet high impact. The University of St. Gallen followed this exact path, and in this book, they proposed practical solutions for enabling sustainable aviation through a unique collaboration between business, aviation and sustainability domains.

The collaboration between different domains within academia, and its consecutive success in yielding practical and accessible solutions to the posed sustainability challenges, also highlights the importance of collaboration between the aviation industry organisations and academia. Working together enabled the University of St. Gallen and SITA for AIRCRAFT to bring two different perspectives on the same issue, exposing conflicting opinions, which then facilitated a collaborative environment of seeking resolutions suitable for both worlds.

We certainly gained new insights from the research done by the University of St. Gallen. It has helped us to develop new ideas and gain an outside perspective in how SITA can tackle industry challenges. We believe that this book provides an excellent and impartial 360-degree view of key issues facing sustainable aviation, and more importantly, it provides ideas and hands-on frameworks on how industry leaders can tackle these issues.

I am personally very grateful to the University of St. Gallen team on this collaboration, especially to Alexander Stauch and Adrian Müller for their patience and openness, to Andreas Wittmer for his support and to Judith L. Walls for teaching me that businesses can and must look at sustainability challenges as an opportunity.

The SITA for AIRCRAFT team that generously provided their expertise and insights to enrich this book are Elham Boozarjomehri, Viktorija Kucerova, Laura Leonardis, Jose De Oliveira and Andrew Yang.

SITA for AIRCRAFT Igor Dimnik
Geneva, Switzerland

Preface

I am, and have always been, in love with aviation. I cannot recall the first time I ever flew—I was a toddler, accompanying my parents on a Mediterranean holiday. My childhood was full of travel, both near and far. As early as I can remember, I had interests in becoming a pilot—dreams that were dashed when I heard about minimum eyesight requirements as a teen, and the carbon footprint of flying in later years.

As kids, my brother and I badgered air hosts and hostesses to let us visit the cockpit at every opportunity, something that was still possible pre-9/11. Once, as a teenager travelling alone, I was invited to join the flight deck during a landing. The shock of cockpit alarms blaring as the plane dipped below a certain altitude created an adrenaline surge through my body. To this day, I do not know if the pilots meant to tease me by omitting to tell me, but upon seeing my shocked face the flight engineer took pity on me and informed me that this was normal. As the pilots guided the giant bird down to earth, it occurred to me that getting 500 tonnes of metal in the air is nothing short of a minor miracle in human and technological ingenuity.

Some of the happiest and saddest moments of my life have been spent on planes and in airports. As a lifelong expat and 'third culture child', flying represented travel for both fun and necessity. My first long-distance travel and re-homing was to South Korea. Things were different back then. For one, people still smoked on planes (I will forever be grateful that those regulations changed). For another, you could not fly directly to South Korea from Europe. The options were to go via Alaska along the northern part of the globe, or via multiple hops along the southern parts of Asia and the Middle East. I was just old enough to become aware of the complex role of international politics in industries like aviation.

I also have fond and less fond memories of airports. I raced through the world's largest airport because of taking the wrong pier towards the gate which was on the opposite end of the airport. I fondly recall a little shack on a dirt runway optimistically labelled 'airport' in someone's elegant cursive hand. I have had to explain why I had a bread knife in my suitcase (thankfully, also pre-9/11), which due to delays ended up travelling with me onboard rather than checked in. I have been stopped by customs for having too many bags, or too few. I was once so sick from food poisoning that I fell asleep right next to the gate and missed my flight. I have gone swimming and watched movies at airports during long layovers and read entire novels on flights. I have research eureka moments in airports and airplanes for lack of having nothing better to do.

In Namibia, I flew in a 4-seater Cessna for the first time, to track predators. My stomach protested strongly at the aerial acrobatics, but it is an experience I will always treasure. I would later also experience helicopter flight—a totally different way of being airborne and one that appeals to my spirit of utter freedom, because you can pretty much go anywhere you want. My imagination has taken me even further afield, into fighter jets and spacecraft featured in movies and science-fiction novels. (And given my stomach's objections to flights with even minimal turbulence, these journeys should probably stay in my imagination).

As I matured, I came to appreciate the business infrastructure that makes flying possible. The incredible international and local collaborations that are required to make an airport run. The training that air traffic controllers, pilots and crew need to undergo, the logistics of shifting millions of pieces of luggage from one plane to another at airports, local transport connections to and from airports, feeding passengers, shopping, border control and immigration and so much more.

Aviation is a deeply embedded part of my existence. Which also means that I represent the troubling statistic that 1% of frequent flyers are responsible for more than half of the greenhouse emissions created by air travel. When exactly I became aware of climate change, I do not recall. It was always on the edges of my knowledge—even in the 1970s, we were already aware that things were going in the wrong direction and that our human and industrial activity were the cause. But it is one thing to know climate change is a problem, and another to link it to your own behaviour.

For at least the past decade, flying has no longer been a romantic adventure for me but rather a guilt-tripped journey full of personal cognitive dissonance, or the discomfort I experience from holding two conflicting values/beliefs. As a truly global citizen, I have friends all over the world. The thought that I may rarely or even never visit them again to avoid creating emissions is a painful one. But with my deep-rooted motivation to preserve our natural environment, flying is increasingly less justifiable. As a sustainability professor, I actively seek (business) solutions to combat environmental problems. As such, the concept of sustainable aviation motivates me, even if I have some reservations about whether the terms 'sustainability' and 'aviation' can ever truly go hand in hand.

It is precisely because of my mixed feelings about aviation and sustainability that I agreed to take part in this book project. Flying has always been my reality, all my life. But now it is time for change. To some extent, that means I (and others) need to change our personal behaviours. These days, I think twice before flying somewhere, and convince peers to let me join events online to reduce my personal footprint: to walk the talk, as it were. But in addition, I believe that we need to create transformation in the aviation industry from a systemic point of view. People are going to want to travel, many of them long distances. So, the puzzle to resolve is how can people travel without creating large footprints?

While the COVID-19 pandemic has put a dent in flying for the last year or so, everything points to air travel continuing to grow in the foreseeable future.

This means the aviation industry faces the challenge of drastically reducing emissions and finding solutions to achieve the goals set out by the 2030 Paris Agreement. Given the ingenuity it takes to get airplanes off the ground, I am confident that the industry can be equally ingenious and creative to move towards net-zero mobility.

These are the questions that each chapter in the book seeks to address, from various perspectives. We were fortunate to have seven master's students in the University of St. Gallen agree to take part in this project. These students worked tirelessly, in less-than-ideal conditions complicated by a global pandemic, to investigate how aviation can become more sustainable. These students' theses—in one case, an award-winning thesis—were the backbone for our project that looked at how technology, consumers, airlines, airports and policy all play a role in moving towards sustainable aviation. In addition, we were fortunate to have the support of SITA as our partner on the aviation business side.

The drive and combination of these young scholars and industry experts gives me some hope that achieving sustainability in aviation is possible. What we need now is political will, business acumen and rapid transformative action to make the possibility a reality.

St. Gallen, Switzerland   Judith L. Walls

Preface

In the aviation industry many talk about the so-called aviation virus. Once you are infected, it is hard to lose it. I was basically infected when I was born. My father, having been a glider pilot and later a motor plane pilot, took me flying from a very young age. I remember one occasion where he asked me to fly the plane by taking the controls—I think I was maybe five or six years old. I was very happy to do so and leaned forward to grab the stick, which I pulled back instantly owing to my short arms. Of course, the plane immediately went into a steep upward climb and my father took over again.

Early on I began to build model planes which I then flew. I had some powered planes, but I was especially fascinated by gliders. So, I carried these model gliders in my backpack up the mountains and flew them in the different winds. Once I was 17 years old, I learned to fly gliders and was lucky to have my father who also flew gliders. Hence, my family spent the weekends at the gliding club. I was eager to compete and managed to enter the Swiss National Junior Gliding Team, where I flew many competitions in Switzerland and abroad—from regional to Swiss national to European championships.

During my studies I travelled the world and got to enjoy international air transport. I backpacked through the world's continents and enjoyed global mobility. Low-cost carriers entered the market and made flying around much cheaper, although in Europe I still preferred interrailing. When I finished my studies, one of my professors told me that I may think about a doctorate degree, which came as a big surprise to me. I was more the practical person, who preferred work with a real impact and could not think about being an academic, as I thought of becoming a pilot. But Swissair had just gone bankrupt then and the airline market was in a crisis after 9/11.

Nevertheless, I was open for any job after my studies and was offered a job as an assistant in the Institute for Tourism and Public Services at the University of St. Gallen. There, I saw that most projects had an impact on the tourism and transport industries and when I was asked to do a project for the Federal Office of Civil Aviation in Switzerland, my motivation went through the roof. It was the first time I was able to combine my passion for aviation with my work. During this period, I got to know the then president of the Swiss Aeroclub organisation, Roland Müller, who was a post-doc student at the University of St. Gallen at the time. We met at a general assembly of the Aeroclub, when I was elected as the President for the Eastern Swiss

Section. During a break we had a good chat and Roland just said that we needed an aviation research centre. I was almost finished with my doctoral studies, and the idea of doing research and applied work for the aviation industry at a university fascinated me. After many talks to professors and the rector, we were given the support to establish the Center for Aviation Competence at the University of St. Gallen. It was set up as an industry-focused centre with the goal to support the aviation industry with research and knowledge to help this complex industry become more successful in a liberalised world.

In my position as Head of the aviation research centre, I joined academic aviation networks and was co-founder of the Swiss Aerospace Cluster, a network consisting of over 160 member companies today. My work at the university let me fly to many places in the world for academic and professional exchanges, and it showed me the great benefits society gains from global air connectivity. In some years I was travelling frequently and reached gold status with two airlines.

Over the years I learnt more about all the benefits but also the costs to society from aviation. Especially newer business models, such as low-cost airlines, gained my interest. Analysing these models shows that by implementing dynamic prices in combination with ancillary sales, airlines can produce very low prices, but still earn a profit. As fixed costs are high, airlines optimise load factors by offering differentiated prices. They know that low anchor prices trigger demand, also if these low advertised prices are never the final costs of a passenger for a flight. Our research has shown that in most cases, the total travel spending of passengers is similar to that of other airlines, only that with low-cost airlines it feels cheaper as many small amounts are paid, which add up to the total flight costs. In the end, many low-cost airlines operating on short haul markets are even more profitable than traditional network airlines.

The question raised in the last few years focused more and more on the topic of sustainability and especially the negative effects, such as emissions from airlines. In this context, it must be questioned if it is right to offer placatively low prices to trigger an overdemand, or in other words a demand which is not really a natural demand, but just a generated demand based on the lowest fares. Many countries have reacted by introducing environmental charges such as CO_2 taxes. The airline industry itself has committed to a global emission offsetting scheme (CORSIA), and there is an agreement to reach a zero CO_2 world in 2050. This target is a huge challenge for the aviation industry and especially for airlines. It may be achieved by new technologies and new CO_2 neutral fuels. But to push new technologies and to scale up the production of CO_2 neutral fuels, huge capital influx is needed. It is hardly possible for the aviation industry to stem these investments itself, and hence, the discussion of CO_2 taxes also brings opportunities, but only if these tax incomes generated can really be used for pushing CO_2 neutrality in aviation. While investing into new technology and fuels, society also must be more aware of the need for travels and especially it has to reflect on prices to be paid for air tickets. Is it right to buy an air ticket for 5 EUR? Have airlines, which offer such prices, understood our dilemma and the problem society runs into? Is it right to create non-natural demand?

Is it ethically right to offer very low base prices and then charge a lot of extra fees, also for services, which travellers do not have a real choice about? And is it right, in times of discussion about the implementation and increase of CO_2 charges, to placatively offer 5 EUR flights?

Questions about how to reduce air travel to what is really needed and about how to achieve CO_2 neutrality as fast as possible are dominant in research. Should airlines be allowed to do marketing communication by stating prices? Should minimum prices be implemented in air transport? Should very short flights of up to 500 km be banned? Should CO_2 charges be higher, the shorter the flight? There are many questions to be raised and answered and I am looking forward to contributing to those in the coming years. I hope you can enjoy reading this book. It was written by two professors and a group of eager PhD and master's students and shows the interest and commitment of the younger generation in helping to find solutions. We all want to enjoy unlimited global mobility. The question is how we can make sure it will be possible in the future.

St. Gallen, Switzerland   Andreas Wittmer

Project Leaders

Alexander Stauch, Institute for Economy and the Environment, University of St. Gallen
Adrian Müller, Center for Aviation Competence, University of St. Gallen

Student Contributors

Philipp Gunziger
Juliette Kettler
Ivan Vuckovic
Nadine Zumsteg
Robin Richner
Annabel Wiegand
Alwin Schmid

Acknowledgements

First of all, we would like to thank Erik Linden who, together with Alexander Stauch, first developed the idea for this book. Erik Linden played an important leadership and communication role, especially at the beginning of the project. Later in the project, he was always available to the researching students with his expertise in the field of aviation. We would like to express our sincere thanks for this commitment.

Secondly, we would like to thank SITA very much. SITA has actively supported us in the project for over a year with practical experience and their general expertise. They specifically networked us with their various experts and were constantly involved in the project. We would like to make a special mention of Igor Diminik, Directory Strategy, Marketing and Sustainability, SITA for AIRCRAFT, as well as Viktorija Kucerova and Elham Boozarjomehri as part of his team. All three of them have kept us going with their passion for aviation and their knowledge of the industry and have been instrumental in shaping the book.

Thirdly, we would also like to thank all academic supervisors for the master's theses. The guidance and support were very valuable for the students on the one hand and ensured high-quality research outcomes on the other hand. Thus, we give special thanks to Karoline Gamma, René Puls and Rolf Wüstenhagen.

Furthermore, we would like to thank dedicated scholars who provided valuable feedback and ideas during the research process. This includes Prof. Glen Dowell from Cornell University and Prof. Rolf Wüstenhagen from the University of St. Gallen.

Finally, we would also like to thank the master's students who did not only continue writing the book chapter after the master's thesis but left us their valuable master's theses as a knowledge base for the book chapters. These students include Robin Richner, Annabel Wiegand and Alwin Schmid.

Last but not least, we would also like to highlight the contribution of Chris Siegrist who did language and spelling checks as well as the language revision of the final manuscript of the book.

About this Book

The original idea for this book project was born in the spring of 2019 at a joint meeting between the Center for Aviation Competence (CFAC-HSG) and the Institute for Economy and the Environment (IWOE-HSG) at the University of St. Gallen. It was clear to both parties that the topic of sustainable aviation will become increasingly relevant in the coming years and that urgent action is needed in the wake of the climate targets that have been set by the Paris Agreement. But what should these actions look like? Which actors should act and how? What is the actual goal of sustainable aviation? This was the moment when the idea of a joint research project with the aim of delivering a book that shapes knowledge and actions in practice towards climate-friendly aviation was born.

The research project "Sustainable Aviation—A Management Perspective" aimed to get to the bottom of these questions together. As the research parties involved are research institutes of a business university, the focus of the project was clearly placed on a practical management perspective with the aim of developing recommendations for action in practice.

In order to explore the many different aspects of sustainable aviation, the first step was to define the relevant main topics and to draw up a rough structure for the book. Based on this structure, seven master's students from the University of St. Gallen were recruited, each of whom worked on one of the relevant main topics as part of a one-year master's thesis.

Fortunately, during the research process, the company SITA could be won as a practical partner, who actively supported the project with its expertise as an advisory body. The students, but also the other main authors of the chapters, always had a lively exchange with SITA, but also with numerous other relevant stakeholders from the industry, which always ensured proximity to practice.

Based on the findings of the seven master's theses, the chapters of the book were then written by different teams of authors. The author teams consisted of master's students, PhD students, post-doc researchers and professors from both institutes involved. The chapters were also supplemented with new content and the book was rounded off with additional intermediate chapters. In the end, the chapters were brought together, and the final book was reviewed by various practitioners and academics before it found its way to publication.

Contents

Sustainable Aviation: An Introduction.......................... 1
Adrian Müller, Judith L. Walls, and Andreas Wittmer

Technology Assessment for Sustainable Aviation................... 23
Alexander Stauch and Adrian Müller

Perceptions of Flight Shame and Consumer Segments in Switzerland... 51
Philipp Gunziger, Andreas Wittmer, and René Puls

Intermezzo: Considerations on the Interdependence of Technology, Consumer Behaviour Change and Policy Interventions to Achieve Sustainable Aviation... 75
Alexander Stauch

Introducing Sustainable Aviation Strategies....................... 91
Judith L. Walls

Airline Perspective.. 109
Juliette Kettler and Judith L. Walls

Controlling, Guiding and Assisting: The Role of Airports in the Transition Towards Environmentally Sustainable Aviation........... 137
Ivan Vuckovic and René Puls

The Role of Public Policy.................................... 163
Nadine Zumsteg and Andreas Wittmer

Towards Sustainable Aviation: Implications for Practice............ 187
Adrian Müller, Alexander Stauch, Judith L. Walls, and Andreas Wittmer

Closing Statement.. 197

About the Editors

Judith Walls is Professor and Chair for Sustainability Management and Co-Director of the Institute for Economy and the Environment (IWÖ), as well as the Delegate of Responsibility & Sustainability at the University of St. Gallen. She researches the intersection of business and environmental sustainability focusing on governance such as behavioural characteristics of boards and managers in the context of stakeholder, shareholder and other institutional pressures on organisations. Her work extends into environmental governance of industries that directly affect land use such as the agriculture, mining and trophy hunting.

Prof. Walls has published in outlets such as *Business & Society*, *European Business Review*, *Journal of Organizational Behavior*, *Journal of Business Ethics*, *Organization & Environment*, *Strategic Management Journal* and *Strategic Organization*. She serves on editorial boards and as guest editor of several journals, and has won many prestigious research awards and grants. Prior to her academic career, Prof. Walls worked in investor relations consulting at Technimetrics/Thomson Financial (now Thomson Reuters) in Europe and the Asia-Pacific. She has also volunteered extensively with conservation organisations in Namibia, Botswana and Mongolia working on topics related to human–wildlife conflict at local community level.

Andreas Wittmer is Head of International Networks, Academic Director of the CEMS Master of International Management and Senior Lecturer in Management with a special focus on aviation at the University of St. Gallen. Furthermore, he is Managing Director of the Center for Aviation Competence and Vice Director at the Institute for Systemic Management and Public Governance. In addition, he holds guest teaching positions at Swiss and international Universities at bachelor's, master's and executive level. His research focuses on network industries, such as aviation. His interests link with a better understanding of efficiencies generated by networks and the impacts on stakeholders. He regularly publishes in international journals and books. He is member of several editorial journal and conference boards. Furthermore, he is Vice President of the Swiss Aerospace Cluster, Vice President of the Swiss Aviation Research Center and freelance aircraft accident investigator at the

Swiss Aircraft Accident Investigation Office. He holds awards for best teaching at Modul University Vienna and the University of St. Gallen, Best Paper Awards from AIEST Conference and Honours from FAI and Pro Aero Foundation for excellent work with the Center for Aviation Competence at the University of St. Gallen.

List of Abbreviations

Aviation Technology Term	Explanation
A-CDM	Airport collaborative decision-making
ACI	Airport Council International
AIC	Aviation-induced cloudiness
APU	Auxiliary power unit
ASSIF	Airport Sustainability Stakeholder Influence Framework
ATC	Air traffic control
ATS	Aircraft towing systems
CO_2	Carbon dioxide
CO	Carbon monoxide
CORSIA	Carbon Offsetting and Reduction Scheme for International Aviation
CEM	Collaborative environmental management
GHG	Greenhouse gas
H_2	Hydrogen
IATA	International Air Transport Association
ICAO	International Civil Aviation Organization
IEA	International Energy Agency
LH_2	Liquid hydrogen
LTO	Landing and take-off
NO_X	Nitrogen oxides
SAF	Sustainable aviation fuel (Biofuel, Synthetic Fuel)
TAW	Tube and wing (classic aircraft design)

List of Figures

Chapter 1

Fig. 1	Development of airfares and demand. Source: Own illustration with data from Bureau of Transportation Statistics, US Department of Transportation	4
Fig. 2	World air traffic flow in 2018 and 2038 (pre-COVID). Source: Airbus (2019)	4
Fig. 3	Influence of external shocks on air transport. Source: Airbus (2019)	5
Fig. 4	The aviation system. Source: Wittmer et al. (2021)	6
Fig. 5	"Mickey Mouse Model". Own illustration based on Peet (2009) ...	11
Fig. 6	Emission scenarios and the resulting radiative forcing levels. Source: Box 2.2, Fig. 1 from IPCC, 2014: Topic 2—Future Climate Changes, Risk, and Impacts. In: Climate Change 2014—Synthesis Report. Contribution of Working Groups I, II, and III to the Fifth Assessment Report of the Intergovernmental Panel on Climate Change [Core Writing Team, R.K. Pachauri and L.A. Meyer (eds.)]. IPCC, Geneva, Switzerland	13
Fig. 7	Key characteristics of the scenarios collected and assessed for WGIII AR5. For all parameters, the tenth to 90th percentile of the scenarios is shown. Source: IPCC (2020) https://ar5-syr.ipcc.ch/topic_pathways.php	14
Fig. 8	GHG emission pathway scenarios. Global greenhouse gas (GHG) emissions (gigatonne of CO_2-equivalent per year, $GtCO_2$-eq/year) in baseline and mitigation scenarios for different long-term concentration levels (**a**) and associated scale-up requirements of low-carbon energy (% of primary energy) for 2030, 2050, and 2100, compared to 2010 levels, in mitigation scenarios (**b**). Source: IPCC (2020). https://ar5-syr.ipcc.ch/topic_pathways.php	15
Fig. 9	Global air passenger traffic forecast comparison, pre- and post-COVID. Source: Air Transport Action Group (ATAG) (2020). Waypoint 2050. https://aviationbenefits.org/media/167187/w2050_full.pdf	19

Chapter 2

Fig. 1	The "Wright 1" Model, a 186-seat all-electric aircraft. Source: Business Traveller (2020)	29
Fig. 2	Energy density based on power-to-weight ratios of different energy sources. Source: Dial (2011)	30
Fig. 3	Airbus ZEROe Hydrogen Aircraft Designs. Source: Airbus Company Homepage (2021)	32
Fig. 4	Hydrogen-powered blended-wing body design by Airbus. Source: Airbus Company Homepage (2021)	33
Fig. 5	New aviation technologies—Investment needed and emissions saved. Own illustration, inspired by Thomson (2020)	42

Chapter 3

Fig. 1	Perception of flight shame. Own illustration	57
Fig. 2	Cluster score distribution (own illustration)	59
Fig. 3	Cluster characteristics (Own illustration)	65
Fig. 4	Sustainability consumer expectations. Own illustration	69

Chapter 4

Fig. 1	Demand decline of cigarettes in Germany from 1991 to 2020. Source: Statista (2021)	79
Fig. 2	Scenarios for high- and low- GHG emissions combined with high and low demand growth in the aviation industry. Own Illustration	82
Fig. 3	Aviation innovations and behaviour change for a sustainable industry transformation. Own Illustration	85

Chapter 5

Fig. 1	CO_2 emissions for International Aviation, 2000–2050 (ICAO, 2019)	93
Fig. 2	The circular economy "value circle". © Zsusza Borsa (2021)	96
Fig. 3	GHG Protocol Scope 1, 2, and 3 emissions. GHG Protocol (2021)	98
Fig. 4	Proportion of GHG emissions from Scope 1 sources—selected industries. Own illustration based on 2018 data from Trucost	99
Fig. 5	Industry's proportion of all Scope 1 (direct) GHG emissions. Own illustration based on 5084 companies' data from Trucost (2018)	100
Fig. 6	Hierarchy of Climate Mitigation Strategies. The most effective strategy is to avoid emissions. Own Illustration	101

List of Figures xxxi

Chapter 6

Fig. 1　Overview of the Four-Pillar Strategy to mitigate climate change. Illustration adapted from IATA (2021) 112
Fig. 2　The funnel metaphor and the ABCD-procedure of the FSSD. Own illustration based on Broman and Robèrt (2017, p. 21) 117
Fig. 3　Visualization of the three sustainability strategies. Own Illustration adapted from Schmidt (2008, p. 14) 118
Fig. 4　Strategy Overview "Green Status Quo." Own illustration. Note. Shaded areas represent efforts on which the strategy focuses. FESA stands for Full Environmental Sustainability for Airlines 126
Fig. 5　Strategy Overview "Multimodal Mobility." Own Illustration. Note. Shaded areas represent efforts on which the strategy focuses. FESA stands for Full Environmental Sustainability for Airlines 127

Chapter 7

Fig. 1　Emission sources per Scope. Adapted from ACI (2009, pp. 15–16) 140
Fig. 2　Singapore Changi GHG emissions scope breakdown in 2017/2018. Adapted from Changi Airport Group (2018, pp. 84–85) 141
Fig. 3　ACI Carbon Accreditation levels. Adapted from Airports Council International (2019, pp. 9, 20) 144
Fig. 4　Airport Sustainability Stakeholder Influence Framework (ASSIF). Own illustration 147

Chapter 8

Fig. 1　Contribution of measures for reducing international aviation Net CO_2 emissions. Source: ICAO (n.d.-b) 166
Fig. 2　Functional Airspace Blocks according to the SES programme. Own illustration 168
Fig. 3　Policy framework to reduce GHG emissions from aviation (own figure) 171
Fig. 4　Triggers for change according to the policy framework. Own illustration 171
Fig. 5　Sustainable aviation according to the policy framework. Own illustration 172
Fig. 6　Considerations for politics' action according to the policy framework. Own illustration 173
Fig. 7　Short-term policy measures to reduce GHG emissions according to the policy framework. Own illustration 174
Fig. 8　Considerations for elaboration of measures according to the policy framework. Own illustration 181

Chapter 9

Fig. 1　Key messages 190

List of Tables

Chapter 1

Table 1	Examples of emission metric values. Own illustration based on Fig. 3.2 and Table 3.1 from IPCC, 2014: Topic 3—Future Pathways for Adaptation, Mitigation, and Sustainable Development. In: Climate Change 2014—Synthesis Report. Contribution of Working Groups I, II, and III to the Fifth Assessment Report of the Intergovernmental Panel on Climate Change [Core Writing Team, R.K. Pachauri and L.A. Meyer (eds.)]. IPCC, Geneva, Switzerland	9
Table 2	CO_2 emissions from commercial aviation, 2018	17

Chapter 2

Table 1	Sustainability evaluation of aviation power and fuel technologies	36
Table 2	Economic evaluation of aviation power and fuel technologies	39
Table 3	Evaluation summary of aviation power and fuel technologies	43

Chapter 3

Table 1	Demographics by segment (own representation)	62

Chapter 5

Table 1	Strategic options available to airlines at the corporate and business level	94
Table 2	Types of corporate climate change strategies	97
Table 3	Hierarchy of climate mitigation strategies including examples	102

Chapter 6

Table 1	Three sustainability strategy approaches for airlines	119
Table 2	Opportunities and threats for SWISS becoming sustainable	124

Chapter 7

Table 1 Overview of the sampled airports 147

Chapter 8

Table 1 Overview of air transport organisations ICAO, EASA, and IATA 166

Sustainable Aviation: An Introduction

Adrian Müller, Judith L. Walls, and Andreas Wittmer

Abstract

- Aviation is a key driver of global economic prosperity which has grown at an average annual rate of 5% in the past and is projected to continue to grow strongly in the future.
- By burning fossil fuels, aviation contributes significantly to greenhouse gas emissions in the transportation sector which are also expected to grow.
- Aviation must mitigate its impact on climate change and adapt its strategies and technologies to achieve the targets the sector agreed upon in the Paris climate agreement.
- Pursuing environmental sustainability should not only seen as a cost but as an opportunity which can be lucrative for first movers who gain strategic advantages.
- Pathways toward sustainable aviation must include all subsystems of the industry and coordinated approaches must account for the interdependencies between different actors.

A. Müller (✉) · A. Wittmer
Center for Aviation Competence, University of St. Gallen, St. Gallen, Switzerland
e-mail: adrian.mueller@unisg.ch; andreas.wittmer@unisg.ch

J. L. Walls
Institute for Economy and the Environment, University of St. Gallen, St. Gallen, Switzerland
e-mail: judith.walls@unisg.ch

© The Author(s), under exclusive license to Springer Nature Switzerland AG 2022
J. L. Walls, A. Wittmer (eds.), *Sustainable Aviation*, Management for Professionals,
https://doi.org/10.1007/978-3-030-90895-9_1

1 Introduction

Mobility is one of the most fundamental and important features of human activity that satisfies the basic and ever itching need to move from one place to another. In essence, people are mobile because they want to make social connections, economic exchanges, or simply broaden their horizons. Once considered a luxury good, mobility has nowadays become a basic commodity, and it is almost impossible to imagine life without it.

In industrialized nations, people have largely unrestricted freedom to travel and a wide variety of affordable modes of transport. If people want to be mobile, they can be. And no other form of transport provides a sense of freedom associated with travel as aviation. It is thanks to aviation that citizens of industrialized, and increasingly developing, nations have been able to explore all four corners of the world with little effort and expense. Aviation has taken on enormous significance in our lives, whether to escape the daily grind of work and immerse ourselves into a relaxed counter-setting or to establish business contacts across the globe. The rapid rise in aviation has born witness to the development of a hyper-mobile, globalized society.

While mobility and aviation undeniably have a positive effect on socio-economic development and global connectivity in society and business, the individual gains in freedom and movement also come at a costly price to the natural environment. Anthropogenic climate change is staring us in the face, prompting us to question our mobility behavior with increasing urgency. As a carbon-intensive industry, aviation is a major contributor to global warming and has been the focus of the global climate strike movements and other concerned stakeholders in recent years.

The truth is that aviation faces a massive paradoxical challenge. On the one hand, aviation is a key driver of global socio-economic development, direct and indirect job creation, tourism, and foreign investment and international trade. On the other hand, aviation has severe negative consequences on our ecological systems as a result of relying on fossil fuels for energy, greenhouse gas emissions that are emitted at high altitude, and negative impacts from infrastructure. These environmental costs are not born directly by the aviation industry, but rather by society and nature at a global level. As a result, the term "sustainable aviation" may seem a bit like an oxymoron.

Yet, in the context of a world that is transitioning towards a more sustainable and circular economy, changing consumer attitudes and behavior, regulatory shifts, and a global pandemic, the aviation industry is facing strong pressure to transform. The question is not whether aviation should take action on sustainability, but rather what does sustainability transformation for aviation entail and how quickly can the aviation industry transform towards sustainability?

Sustainability of the aviation industry is, therefore, a critical issue in the provision of mobility. In this book, we not only highlight the problems aviation causes, but also actively seek constructive solutions towards a more sustainability-oriented form of aviation. We discuss different aspects of the aviation industry and also a range of different solutions across stakeholder groups. There is likely no silver bullet that can fix to solve the sustainability problems of the aviation industry. But by removing the blinders to take a brutally honest look at the sustainability issues in the industry, and by considering the role of consumers, regulators, operators, we come closer to finding workable solution for sustainable aviation.

2 Global Aviation: An Overview

2.1 Development and Status Quo

Aviation is an important industry for the economy and society as it is a global enabler for international trade and social exchange. As a key driver of globalization, it significantly contributes to economic prosperity. Before the COVID-19 pandemic, air transport was estimated to be worth over USD 1.3 billion of which USD 1 billion are direct, indirect, and induced effects. Pre-COVID, the industry employed about 65.5 million people and transported roughly 4.5 billion passengers per year (ATAG, 2020). In addition to passenger transport, aviation is a key factor in global supply chains with more than 0.9 billion tonnes of freight transported annually (ATAG, 2020). While this represents less than 1% of the total global trade volume, it accounts for about 35% of the *value* in global trade (Shepherd et al., 2016). Air transport is an integral constituent of human connectivity, being indispensable not only for leisure and business travel but also for global economic integration (Conrady et al., 2019).

There are few other industries that have experienced such strong growth as aviation in recent decades. Following the worldwide deregulation of the industry, airlines entered fierce price competition causing ticket prices to decline 22% on average between 1978 and 1993 (Morrison & Winston, 1997). As a consequence, passengers were able to benefit from lower prices, making air transport available to a much larger proportion of the population. In addition to deregulation, technological advances such as the introduction of large capacity aircrafts and, at a later stage, IT solutions (reservation management, online reservation systems) further accelerated the drop in airline prices, fueling the strong growth. Most recently, with the introduction of low-cost carriers (LCC), air transport became a cheap and popular form of transport on short-haul routes. In short, aviation has seen a decreasing trend in ticket prices and a stark growth in passengers over the past few decades (Wittmer et al., 2021). This trend is illustrated in Fig. 1 using the example of the U.S. domestic market. The global economy, measured by gross domestic product (GDP), has grown by 2.8% annually since 1995, while global passenger air traffic, measured in revenue passenger kilometers, has grown at an average annual rate of 5.0%.

While these figures are certainly important to understand how the industry developed, in the context of sustainable aviation, it is more important to look ahead at how the sector is expected to advance in the future. While forecasts always involve an element of the unknown and should be interpreted with care, forecasts by major Original Equipment Manufacturers (OEMs) like Airbus and Boeing can nevertheless provide interesting and important insights. In Fig. 2, Airbus compares the current air traffic volume (as of 2018) with their forecast for the year 2038. What can be seen is that the growth rates in Asia-Pacific and the Middle East are expected

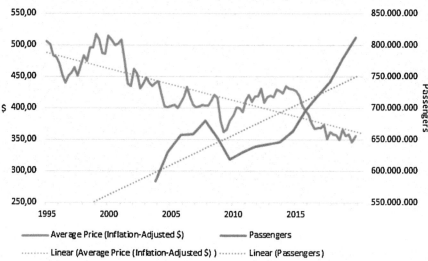

Fig. 1 Development of airfares and demand. Source: Own illustration with data from Bureau of Transportation Statistics, US Department of Transportation

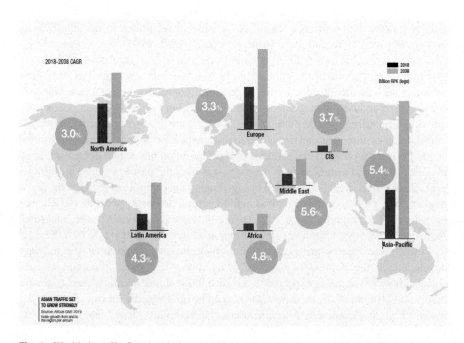

Fig. 2 World air traffic flow in 2018 and 2038 (pre-COVID). Source: Airbus (2019)

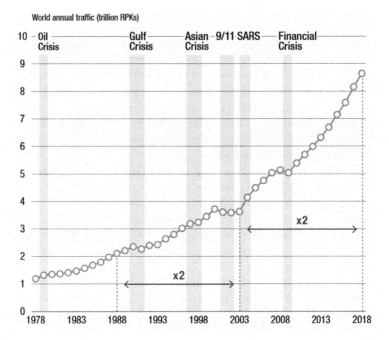

Fig. 3 Influence of external shocks on air transport. Source: Airbus (2019)

to cause a major shift in the air transport market. While in the past the US domestic and European market accounted for a majority of global flights, in the future over 50% of the world's air traffic flows will originate in Asia. All regions of the world, however, are expected to continue to show a positive growth rate in terms of the number of revenue passenger kilometers (RPK).

The skies have not always been blue for aviation, however, and aviation lives up to its reputation as a VUCA industry (volatile, uncertain, complex, ambiguous). In the last 15 years, the civil aviation development has been severely affected by crises that were directly or indirectly aviation related. The Asian crisis in 1998, the terrorist attack in the USA on September 11, 2001, the outbreak of severe acute respiratory syndrome (SARS) in 2003, the world financial crisis of 2008–2009, and most recently the COVID-19 pandemic have all had a negative impact on the overall profitability of the aviation system. The COVID-19 crisis is on an unprecedented scale, the aftermath of which was still ongoing at the time of writing. History has shown that world aviation has always recovered from crises (ICAO, 2020, online) (see Fig. 3). Section 5 in this chapter discusses, what the COVID-19 crisis means for the industry and the path towards more sustainable aviation.

2.2 The Aviation System and Its Stakeholders

Aviation is an industry consisting of various subsystems with countless stakeholders and complex interdependencies. It includes the five subsystems (political, technological, economic, social, and ecological) which influence the stakeholders in the market (Wittmer et al., 2021), which we represent conceptually in Fig. 4. The goal is to give an overview of the most important actors in the model and how they and the respective subsystems influence each other. By briefly introducing each subsystem, and the interdependencies among the different perspectives, we also highlight their influence on the issue of sustainability.

The *technological subsystem* deals with the technological possibilities that set the framework for everything that is feasible in aviation, from basic aircraft design to decarbonization methods. The aviation industry has always been strongly dependent on technological developments, such as aerodynamics or propulsion technology, innovations which not only have a major impact on economic development but also on the ecological impact of the sector. This subsystem represents a major influencing factor for numerous stakeholders in the air transport market. Every actor in the supply system is somehow affected by technology. Most obviously, large OEMs such as Boeing and Airbus drive technological advancement in the consolidated aircraft market. While their product development focus follows more of an outside-in approach, with airlines largely determining the OEM's production with their orders, many other innovative technical approaches stem from an inside-out approach, often by start-up companies. These are regularly spin-offs from the engineering departments of universities, further developing scientific findings from this other important player in the technological subsystem. Sustainable technology

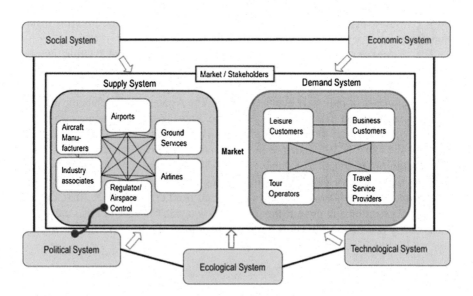

Fig. 4 The aviation system. Source: Wittmer et al. (2021)

development is, therefore, driven by both these approaches, but also influenced by insights from other industries such as automotive. The rapid growth of airlines is due in no small part to technological advances. Digitalization has driven distribution in particular, but other areas such as operations and revenue management have also been significantly influenced by technological innovations. The same can be said about airports, ground services, and airspace control. Especially the air navigation service providers (ANSP) would not be able to handle today's air traffic volume without specialized IT systems.

In aviation, as a highly regulated industry, technological innovations are not always readily applicable due to specific rules and recommendations. This not only slows down innovation processes, but often makes them more costly. The *political/regulatory subsystem* is one of the central reasons for the high complexity of the aviation system, both on national and global level. The subsystem is mainly concerned with issues such as safety and liberalization, which again, have a substantial influence on the sustainable development of the industry. Compliance with the respective regulations is a prerequisite for all stakeholders and for any aviation activities. Aviation legislation is regulated at various levels. Under international law, the ICAO sets the framework for global aviation. Regionally, authorities such as the FAA (USA) or EASA (EU) are responsible for the relevant aviation regulations. In addition, aviation is also the plaything of politics. Especially when it comes to sustainability, it is obvious that aviation has to face complex policy discourses from international to local level. It is not uncommon for the courts to ultimately have to juxtapose national concerns and individual interests of local residents. The great involvement of politics is the result of the social and economic importance of aviation. These two subsystems are very closely linked.

The *economic subsystem* is concerned with the contribution of aviation to the prosperity of nations and is highly important for global economic development. The economic benefits of air transport can be divided into direct, indirect, induced and catalytic effects, which represent to a large extent the income and jobs generated by the aviation system. Estimated economic effects are 10.2 million jobs and USD 704 billion (global GDP) as direct contribution, 10.8 million jobs and USD 638 billion as indirect contribution, 7.8 million jobs, and USD 454 billion and another 36.7 million jobs worldwide by induced effects (ATAG, 2020; Wittmer et al., 2021). If you look at these figures, it becomes obvious how important aviation is and how large its share in prosperity is estimated to be. On the other hand, it also shows how large the group of stakeholders with corresponding demands and expectations is.

However, it is not only economically that aviation has a great influence on society. The *social subsystem* is concerned with these impacts. Economic opportunities enhance social welfare at the local and regional level and positively influence people's standards of living and lifestyles. In this context, mobility is viewed as a factor that enhances the quality of life and opens up access to different countries. As such, the industry serves as a means of mass transportation, which helps to connect different countries and cultures. This also fosters cultural awareness and increases multicultural collaboration. In addition, aviation enables people to reach places that would otherwise be inaccessible and can also supply necessary goods and services. The negative aspects include the effects of noise on the well-being of people living in the vicinity of airports. Noise may have several health effects, both medical and psychological, on individuals (Wittmer et al., 2021).

Finally, and this is the core focus of the book, aviation has an impact on the natural environment. The *ecological subsystem* is mainly negatively affected by aviation, in contrast to the other subsystems. The global environmental impacts of aviation include the effects on the atmosphere and thus on the entire world. Regional and local impacts refer to the environmental effects at particular airports and the adjacent communities. Emissions are mainly produced by the aircraft engine combustion process, which burns a blend of a variety of hydrocarbons. Section 4 in this chapter looks at aviation's emissions and contributions to climate change in detail.

In addition, the conceptual model also represents the aviation market, where supply and demand come together. Here it is important to emphasize that demand is very heterogeneous. Private and corporate customers book air travel for leisure or business purposes and to fulfill a myriad of different needs. In turn, there are various interrelationships, so that most aviation customers are also affected by aviation in other subsystems. Chapter "Perceptions of Flight Shame and Consumer Segments in Switzerland" on consumer behavior shows the impact of these complex relationships on demand for sustainable aviation.

We can see that due to the multidimensional nature of aviation, there is a close interconnectedness between political, technological, economic, social, and environmental issues. While any of these perspectives can bring valuable understanding to the field of aviation by itself, the complexity and interdependencies between the different aspects require taking a more integrated approach on the sustainable development of the industry. With this book we aim to take into account these interdependencies in order to show possible ways forward for managers and policy makers, who are tasked to make aviation more sustainable.

3 Climate Change and Environmental Sustainability

3.1 Definitions and Causes

When analyzing a complex issue such as sustainable aviation, it is crucial to first define two key terms and concepts: climate change and sustainability.

3.1.1 Climate Change

As a starting point, it is imperative to understand that climate and weather are not synonymous. While weather focuses on a period of days, or at most, weeks, *climate* describes the atmospheric conditions over a time horizon of three decades (Seinfeld & Pandis, 2006).

The definition of climate change comes from two key sources: the Intergovernmental Panel on Climate Change (IPCC) and the United Nations Framework Convention on Climate Change (UNFCCC). The UNFCCC defines climate change as "a change of climate which is attributed directly or indirectly to human activity that

Table 1 Examples of emission metric values. Own illustration based on Fig. 3.2 and Table 3.1 from IPCC, 2014: Topic 3—Future Pathways for Adaptation, Mitigation, and Sustainable Development. In: Climate Change 2014—Synthesis Report. Contribution of Working Groups I, II, and III to the Fifth Assessment Report of the Intergovernmental Panel on Climate Change [Core Writing Team, R.K. Pachauri and L.A. Meyer (eds.)]. IPCC, Geneva, Switzerland

	Lifetime (years)	GWP		GTP	
		Cumulative forcing over 20 years	Cumulative forcing over 100 years	Temperature change after 20 years	Temperature change after 100 years
CO_2	n.a.	1	1	1	1
CH_4	12.4	84	28	67	4
N_2O	121.0	264	265	277	234
CF_4	50,000.0	4880	6630	5270	8040
HCF-152a	1.5	506	138	174	19

alters the composition of the global atmosphere and which is in addition to natural climate variability observed over comparable time periods" (United Nations [UN], 1992, p. 19).

While the UNFCCC's definition focuses on anthropogenic (human-induced) change, the IPCC additionally includes climate change resulting from natural causes. In this context it is important to note that while there has always been a certain natural variability in climate, it is the scale of anthropogenic impact on climate change that is considered to be the critical driver of the current climate change problem. For example, the average global temperature has varied in the last 1000 years without anthropogenic influences. However, by comparison, the twentieth century is recorded as the warmest century in history, and the three decades leading up to 2012 were "the warmest 30-year period of the last 1400 years" (IPCC, 2013) in the northern hemisphere.

The main causes of anthropogenic climate change are greenhouse gases (GHGs) which are released in the atmosphere. The most important GHGs include carbon dioxide (CO_2), methane (CH_4), nitrous oxide (N_2O), hydrofluorocarbons (HFCs), perfluorocarbons (PFCs), and sulfur hexafluoride (SF_6) (UN, 1998). When the concentration of these gases increases, they contribute to the so-called radiative forcing, which means they increasingly hinder energy entering the earth's atmosphere from radiating back into space. The effectiveness of GHGs to cause radiative forcing is expressed by their global warming potential (GWP), which translates the radiative forcing caused by a specific GHG into a CO_2-equivalent (CO_2-eq). While all GHG have a considerably higher GWP than carbon dioxide, CO_2 is responsible for most of GHG emissions (IPCC, 1996). In 2010, CO_2 caused over two-thirds of the anthropogenic GHG emissions (expressed in gigatons of CO_2-eq) (IPCC, 2014). Table 1 shows examples of emission metric values.

In addition, climate change is intertwined with other major environmental sustainability problems particularly biodiversity loss in a self-reinforcing feedback loop. For example, climate change exacerbates the frequency, intensity, and scope of wildlife fires, habitat and land use changes, drought and rain patterns, and species extinction that all affect biodiversity (Lovejoy, 2006). In turn, biodiversity loss results in less resilience of the Earth's ecosystems and subsequent lower ability of the ecosystem to absorb carbon and other ways to mitigate climate change.

3.1.2 Sustainability

The mostly commonly cited definition is from the influential Brundtland report, which defines sustainable development as "development that meets the needs of the present without compromising the ability of future generations to meet their own needs" (WCED, 1987). This report also referred the three aspects of sustainability as the economic, societal, and ecological systems. As a result, in the business and industry context, sustainability is often defined as the "triple bottom line" in which economic success, ecological responsibility, and social justice go hand in hand (Elkington, 1997).

A common critique of the triple bottom line view of sustainability, however, is that it encourages a weak form of sustainability, in which the social and ecological considerations are often traded off against economic development (Elkington, 2018). As a result, business often ends up decoupling sustainability goals from economic ones, in a form of greenwashing that is sometimes also called the "Mickey Mouse model" (Peet, 2009). By contrast, a strong definition of sustainability is one in which the economic system is nested within, and depends on, the social and ecological systems (Giddings et al., 2002). This definition explicitly acknowledges the interconnected and interdependent relationships between human industrial activity, society, and nature (Fig. 5).

From a strong sustainability perspective, it is clear that to preserve our closed ecological system, we need to replace non-renewable resources with renewable ones and that our use of renewable resources must be at a pace no faster than at which they can naturally regenerate. Moreover, an intact and functioning ecosystem is required so that future generations can meet their needs. A warming and unstable climate is, therefore, considered unsustainable.

While the economic and social aspects of sustainability are certainly relevant in the aviation context, the focus on this book lies on the environmental dimension, by which we mean the Earth's ecological system. To avoid irreversible damage to the earth's ecosystem, we must take action to reduce GHG emissions before 2050. Without measures to combat climate change, global temperatures in 2100 could increase by 4.5 °C compared to pre-industrial times, which would result in catastrophic consequences for our planet (IPCC, 2014).

While the economic dimension of sustainability cannot be completely set aside in a highly competitive industry like aviation, it is clear that if the aviation industry fails to address environmental sustainability, the industry will face much more severe consequences in the future, such as a rising price on carbon, regulatory interventions, and legitimacy of the industry. Like other industries, aviation has little choice but to mitigate its impact on climate change and adapt its strategies and technologies. However, pursuing environmental sustainability strategies should not be viewed only as a cost; such new approaches also provide opportunities and, in some cases,

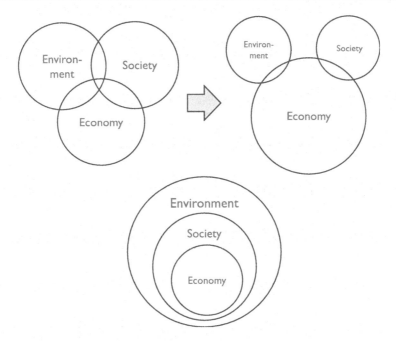

Fig. 5 "Mickey Mouse Model". Own illustration based on Peet (2009)

can be quite lucrative. Those that move first will benefit competitively and gain strategic advantage by staying ahead of peers and regulators, and avoiding severe downside consequences by not addressing climate change (Hart, 1995; Porter & Kramer, 2006). Our aim in this book is to highlight where potential opportunities lie, and what aspects represent the biggest threats to the future viability of the industry, and how these can be overcome by embedding sustainability into the aviation industry.

3.2 Consequences and Time Horizon

Time is a critical factor in combating climate change. The emitted GHGs remain in the atmosphere for decades and thus also influence radiative forcing for many years. Therefore, past emissions alone are likely to cause 1.5 °C warming compared to pre-industrial times. If current trends continue, emissions will cause additional warming, resulting in sometimes irreversible damage to the Earth's ecosystems (IPCC, 2018). The consequences are manifold and range from melting of the glaciers, global sea-level rise, warming and acidification of oceans, and heavy precipitation, to temperature extremes, an increase in droughts and species loss, just to name a few (IPCC, 2018).

Most vulnerable to these effects are developing countries which are generally more dependent on the primary sector (agriculture) and have less robust infrastructure to withstand extreme weather. Estimates assume that the costs of a natural disaster can amount up to 5% of the GDP of a low-income country (Stern, 2007). The Stern report further estimates, in an attempt to quantify the total cost of climate change, that climate change risks will be equivalent to an annual loss of between 5–20% of global GDP. This is disproportionate to the estimated 1% of global yearly GDP estimated for the measures needed to avoid the worst consequences of climate change. The associated loss in biodiversity, reinforced by climate change, is estimated to cost business and industry USD 5 trillion per year, or USD 150 trillion in total, twice the global GDP (Kurth et al., 2021).

Since the Paris Agreement of 2015, there is broad worldwide consensus among researchers, experts, and politicians that the temperature rise needs to be kept as low as possible. In a supplement to the UNFCCC, the nations' participating in the summit agreed to significantly reduce their GHG emissions in order to halt climate change before an irreversible tipping point is reached. For this purpose, all participating nations have submitted their nationally determined contributions including measures to avoid, adapt to, or cope with climate change (Rogelj et al., 2016).

The main overarching climate goal of the agreement is the following: "Holding the increase in the global average temperature to well below 2 °C above pre-industrial levels and to pursue efforts to limit the temperature increase to 1.5 °C above pre-industrial levels" (UNFCCC, 2015). To achieve this goal, concrete, rapid and, above all, effective measures are essential in all sectors. The transportation sector will have to play a central role.

Measures to drastically reduce GHG emissions must come into effect very soon if warming is to be kept below 2 degrees. This realization is important for aviation insofar as the innovation and approval cycles can sometimes take a long time. Aviation cannot wait until, for example, promised technologies (see also Chap. "Technology Assessment for Sustainable Aviation") are ready for use. Measures to reduce emissions must be taken and implemented immediately. The IPCC has developed the Representative Concentration Pathways (RCP) which make it understandable how large the global temperature increase between 2010 and 2100, caused by a certain amount of emitted CO_2-eq (the so-called carbon budgets), would be (Fig. 6).

The RCP include a so-called baseline scenario (a future without actions to reduce emissions) as well as several pathways to limit global warming below 2 °C. Figure 7 shows model pathways and what a change in CO_2-eq emissions is likely to result in temperature rise. It can be seen that in order to not overshoot the 2° temperature level, scenario RCP2 should be achieved, which requires a significant reduction of CO_2-eq emissions already by 2050. This emphasizes the need for the aviation industry to take immediate action. The longer we are on the trajectory towards the wrong RCP scenario, the more difficult it will become and the more severe the measures will need to be to reach the target scenario RCP 2.6.

Figure 8 shows all RCP scenarios and what they mean for the "allowable" annual GHG emissions and how this affects the necessary upscaling in low-carbon energy sources. To achieve RCP 2.6, a 310% increase in low-carbon energy share of

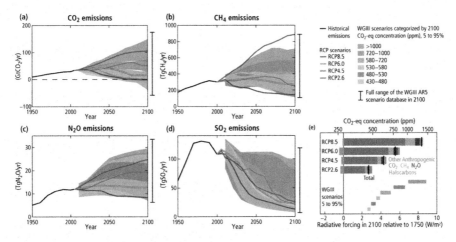

Fig. 6 Emission scenarios and the resulting radiative forcing levels. Source: Box 2.2, Fig. 1 from IPCC, 2014: Topic 2—Future Climate Changes, Risk, and Impacts. In: Climate Change 2014— Synthesis Report. Contribution of Working Groups I, II, and III to the Fifth Assessment Report of the Intergovernmental Panel on Climate Change [Core Writing Team, R.K. Pachauri and L.A. Meyer (eds.)]. IPCC, Geneva, Switzerland

primary energy by 2050 based on 2010 levels is required, again providing evidence that measures in the transportation sector are urgent.

4 Aviation and Climate Change

4.1 Aviation's Emission Sources

Section 2.2 in this chapter shows that aviation is an industry with numerous heterogenous stakeholders and complex processes. Therefore, it is difficult to identify all emission sources of the sector. To facilitate the categorization, the World Business Council for Sustainable Development (WBCSD) and the World Resources Institute (WRI) use the concept of scopes, which structures the broad spectrum of organizational impacts on the GHG emissions (WBCSD & WRI, 2004). Several actors from the industry are making use of this concept when reporting their CO_2 emissions.

- *Scope 1*: direct emissions caused by "sources that are owned or controlled by the company" (WBCSD & WRI, 2004). E.g. for an airline this includes emissions from burning fossil fuel in flight operations and own ground vehicle traffic, and if applicable, own energy production.
- *Scope 2*: indirect emissions, e.g. production of purchased energy (WBCSD & WRI).
- *Scope 3*: the rest of indirectly caused emissions, i.e. all other emissions which result of the company's activities, but not owned by the company (WBCSD &

CO$_2$-eq Concentrations in 2100 (CO$_2$-eq) Category label	Subcategories	Relative position of the RCPs	Change in CO$_2$-eq emissions compared to 2010 (in %)^c		Likelihood of staying below a specific temperature level over the 21st century (relative to 1850-1900)			
			2050	2100	1.5°C	2°C	3°C	4°C
< 430	Only a limited number of individual model studies have explored levels below 430 ppm CO$_2$-eq^j							
450 (430 – 480)	Total range	RCP2.6	-72 to -41	-118 to -78	More unlikely than likely	Likely		
500 (480 – 530)	No overshoot of 530 ppm CO$_2$-eq		-57 to -42	-107 to -73		More likely than not		
	Overshoot of 530 ppm CO$_2$-eq		-55 to -25	-114 to -90		About as likely as not		
550 (530 – 580)	No overshoot of 580 ppm CO$_2$-eq		-47 to -19	-81 to -59		Unlikely	Likely	Likely
	Overshoot of 580 ppm CO$_2$-eq		-16 to 7	-183 to -86		More unlikely than likely^9		
(580 - 650)	Total range	RCP4.5	-38 to 24	-134 to -50				
(650 - 720)	Total range		-11 to 17	-54 to -21		Unlikely	More likely than not	
(720 - 1000)^b	Total range	RCP6.0	18 to 54	-7 to 72		Unlikely^k	More unlikely than likely	
> 1000^b	Total range	RCP8.5	52 to 95	74 to 178		Unlikely^k	Unlikely	More unlikely than likely

Fig. 7 Key characteristics of the scenarios collected and assessed for WGIII AR5. For all parameters, the tenth to 90th percentile of the scenarios is shown. Source: IPCC (2020) https://ar5-syr.ipcc.ch/topic_pathways.php

WRI, 2004, p. 25). E.g. for an airline this entails emissions from the supply chain (kerosene, aircraft manufacturing, etc.)

As Scope 1 emissions are easiest to grasp, Scope 1 also gets the most attention in industry. In the case of aviation this also explains why the sustainability discussion is mostly about airlines and their burning of fossil fuels, which is responsible for the majority of their direct emissions and the resulting effects.

Looking at this in more detail, carbon dioxide (CO$_2$) is the best-known pollutant produced by this combustion process through the reaction of carbon oxide (CO) and oxygen in the air (O$_2$). CO$_2$ is a trace gas with a long residence time in the atmosphere (about 100 years). After the emission by airplanes, it is distributed relatively evenly in the atmosphere.

Other gases being emitted to a lesser extent are nitrogen oxide (NO) and nitrogen dioxide (NO$_2$), which together form NO$_x$. NO$_x$ is a greenhouse gas that simultaneously causes the formation of ozone (O$_3$) and the breakdown of methane (CH$_4$) in the atmosphere. Ozone is a greenhouse gas that has both positive and negative

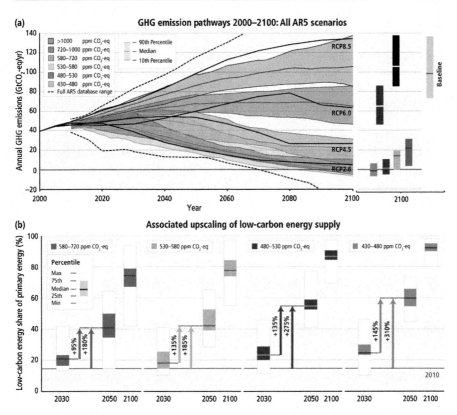

Fig. 8 GHG emission pathway scenarios. Global greenhouse gas (GHG) emissions (gigatonne of CO_2-equivalent per year, $GtCO_2$-eq/year) in baseline and mitigation scenarios for different long-term concentration levels (**a**) and associated scale-up requirements of low-carbon energy (% of primary energy) for 2030, 2050, and 2100, compared to 2010 levels, in mitigation scenarios (**b**). Source: IPCC (2020). https://ar5-syr.ipcc.ch/topic_pathways.php

ramifications. On the one hand, in the stratosphere, it helps filter out harmful ultraviolet (UV) radiation from the sun, thus protecting terrestrial life. On the other hand, an elevated concentration of ozone in the troposphere enhances the greenhouse effect.

Another relatively short-lived greenhouse gas emitted by aircraft, which usually disappears within 1–2 weeks, is water vapor (H_2O). When emitted into the stratosphere, water vapor can help deplete ozone and thus contribute to global warming. Condensation trails are linear ice clouds that form from water emitted by aircraft, preferably in cold and humid air in the upper troposphere and contribute to global warming due to induced radiative forcing. NO_x, water vapor, and other emitted particles have the greatest concentration near their emission source. Combustion of 1 kg of kerosene and 3.4 kg of oxygen typically produces 3.15 kg of carbon dioxide (CO_2) and 1.24 kg of water vapor (H_2O). Depending on engine design and characteristics, 6–20 g of nitrogen oxides (NO_x), 0.7–2.5 g of carbon monoxide

(CO), 01–07 g of unburned hydrocarbons (C_xH_x), and 0.01–0.03 g of soot are also emitted (Wittmer et al., 2021).

Complicating matters for the aviation industry, emissions released in higher altitudes contribute more to climate change than the ones released on the ground (c.f. Jungbluth & Meili, 2019; Umweltbundesamt, 2019; Wittmer et al., 2021). In particular, the GWP of GHGs other than CO_2 is higher at higher altitudes. In addition, NO_x and water vapor affect the atmosphere differently depending on altitude and weather (Cairns & Newson, 2006).

However, it is not enough to consider only the fuel emissions of airlines when discussing sustainable aviation. For example, even though airlines mostly focus on the easier to control Scope 1 and 2 emissions, they also try to influence Scope 3 emissions. However, as the largest potential of mitigating climate change in the industry is rooted in Scope 1 emissions, the majority of the discussion in this book will focus on direct emissions from aviation.

4.2 Aviation's Contribution to Climate Change

Since 1970, direct greenhouse gas emissions from the transport sector have increased by more than 250% and global mobility volumes will continue to grow. Without aggressive and sustained mitigation policies, emissions from transportation could grow faster than emissions from other energy sectors (Sims et al., 2014).

Transportation accounts for around 24% of global CO_2 emissions from energy. Road travel therein is responsible for almost 75% of the sector's emissions. Aviation—despite getting a large portion of the attention in discussions about climate change action—accounts for only 11.6% of transport emissions (IEA, 2020). Nevertheless, we believe that aviation will have a significant role to play in the decarbonization of mobility. For one thing, no other subsector is as strongly associated with global trade, progress, and prosperity as aviation. For another, this industry, which still holds a special fascination, is showing unparalleled growth, which is in stark contrast to climate targets. Therefore, the industry has a signal effect: if aviation succeeds in achieving the climate targets without endangering prosperity, any industry can do so. For this reason, this book is dedicated to the topic of "Sustainable Aviation."

To get an understanding of the problem, we will first take a look at some numbers. Per year, air transport emits around 1 billion tonnes of CO_2 per year. This accounts for approximately 2.5% of total global emissions. International shipping contributes a similar amount, at 10.6% of transport emissions. In comparison, emissions from rail account only for about 1% of all transport emissions. To put this into perspective, Canada and Germany each contributes approximately 2% to global CO_2 emissions.

However, experts do not fully agree on the actual global warming potential of aviation. Jungbluth and Meili (2019) estimate that to account for the total GWP, direct CO_2 emissions of the aircraft should be multiplied by a radiative forcing index (RFI) of 2 on total aircraft CO_2 (or 5.2 for the CO_2 emissions in the higher atmosphere). Cox and Althaus (2019) recommend a factor of 2 to multiply the

Table 2 CO_2 emissions from commercial aviation, 2018

Rank	Country	Per capita CO_2 emissions from international aviation (kg)
1	Iceland	3505.6
2	Qatar	2472.7
3	United Arab Emirates	2195.1
4	Singapore	1741.0
5	Malta	991.6
6	New Zealand	640.3
7	Mauritius	599.8
8	Ireland	574.1
9	**Switzerland**	**513.3**
10	Australia	495.9
11	Bahrain	483.1
12	Netherlands	467.3
13	Panama	446.9
14	United Kingdom	422.1
15	Denmark	413.0

Source: Graver, B., Zhang, K., & Rutherford, D. (2019). The International Council of Clean Transportation. https://theicct.org/publications/CO2-emissions-commercial-aviation-2018

emissions during cruise phase (>9000 m) and then to add this to the CO_2 emissions of the other flight phases to estimate the total GWP.

In addition, the figure of 2.5% total global emissions is somewhat misleading because global CO_2 emissions from aviation are very unevenly distributed among countries, making the average figure not particularly meaningful. It is estimated that in 2018, only between 2–4% of the global population flew internationally. As a result, half of the worlds' commercial aviation emissions are emitted by only 1% of the world's population (Gössling & Humpe, 2020).

These differences not only exist between frequent flyers and non-frequent flyers but also between countries. For example, in Switzerland, a country whose per capita CO_2 emissions from international air travel are among the highest in the world (Table 2), international flights correspond to around 10% of the total CO_2 Swiss emissions or 18% of the CO_2 emissions caused by the transport sector (Swiss Federal Office of Civil Aviation [FOCA], 2021). It should be noted that Switzerland, as a small, sparsely populated and landlocked country in the heart of Europe, has no significant domestic air traffic and, at least continentally, does have alternatives to air travel. A strong export economy and prosperity nonetheless result in high per capita emissions from aviation, thereby already highlighting some of the complex socio-economic interdependencies we encounter when examining the issue.

The importance of the transportation sector in general and aviation in particular, in combating climate change has been recognized by a wide range of stakeholders, as reflected in various developments in society, technology, and policy. The unprecedented boom in electromobility or the recent comeback of night trains in Europe are preliminary signs that the mobility sector is beginning to change. If we now look specifically at the aviation sector again, changes are also evident. For example, regulators at national and supranational level are discussing the introduction of climate levies for aviation or minimum prices for airline tickets. OEMs and startups alike are researching possibilities for climate-neutral aviation, be it biofuels, electric propulsion systems or completely new aircraft concepts. The industry itself has also recognized that sustainability will be the most important strategic issue for aviation in the long term. With the Carbon Offsetting and Reduction Scheme for International Aviation (CORSIA), developed by ICAO, the industry was able to agree on uniform and mandatory climate protection measures for the first time. In concrete terms, the goal is climate-neutral growth from 2020 (ICAO, 2016). A growing societal interest in the issue has undoubtedly contributed to these developments, culminating in considerable public pressure. The climate strike movement with its so-called Fridays for Future protests gave the issue a strong and most importantly loud and visible boost. As a result, media interest has also risen sharply, and aviation has moved into the public spotlight.

4.3   Impact of Climate Change on Aviation

One important finding is that aviation not only contributes to climate change, but is also directly and sometimes severely affected by its impacts. Ryley et al. (2020) present important findings of scientific studies on the impacts of climate change on the aviation system. The biggest problem for aviation concerns changes in the weather. More severe and frequent weather events have a far-reaching impact on airlines, airports and thus other stakeholders. Airlines face challenges in safety, routing, and weather forecasting, and operational adjustments to routes to avoid severe turbulence and pilots needing additional training in the event of lightning strikes. Moreover, strong short duration storms can lead to airport closures, which in turn has far-reaching effects on airlines, passengers, and air traffic control. In the long run this means that actors in the aviation system have to re-assess their vulnerability to climate change and identify potential impacts on operations, safety, and finances. Risk assessments include the evaluation of probability and severity of the impact—both variables that are affected by climate change. On a strategic planning level, this has implications, for example, for airport planning and expansion projects. Airports at sea-level face the threat of rising sea levels and need to adapt. As a whole, the system needs to prepare for networks which become increasingly susceptible to disturbances caused by the weather. In the end this will become a topic also for regulatory bodies, policymakers, and transportation infrastructure planning in general.

It can, therefore, be said that aviation has an inherent interest in actively contributing to the mitigation of climate change in the sense of risk prevention before the entire industry is endangered.

5 Impact of COVID-19 on Sustainable Aviation

As briefly touched upon earlier, the COVID-19 pandemic has been an external shock with unprecedented ramifications for aviation. In 2020, ICAO estimates that international passenger traffic has plummeted by over 50% (ICAO, 2021). While the sustainability issue was at the top of the list of priorities for airline executives before the crisis, these have been superseded by existential issues, at least temporarily. In this book, the topic of sustainable aviation is dealt with independently of COVID-19. On the one hand, we are convinced that the pandemic will pass and that the climate crisis will be the more important strategic topic in the medium to long term. Secondly, we assume that the impact of the pandemic will not significantly change the starting position of aviation in the context of climate protection.

Various scenarios exist for future global passenger volumes. In a medium scenario, it is assumed that the predicted RPK in 2050 will be 16% below the pre-crisis forecast (Fig. 9). This represents neither a decline nor a reversal of the growth trend—at most, it represents a slowdown in growth, which will still not be sufficient to meet aviation's climate targets without further action.

For this reason, the topic of sustainability has by no means lost its importance for aviation. The perception of the importance of effective climate protection measures has been heightened by the pandemic. At the same time, however, the complexity of the topic is also increasing. Financial losses by airlines could jeopardize the available investment capital for decarbonization measures and fleet renewal, GDP losses could reduce the willingness of politicians to take effective policy measures, and a social pent-up demand could additionally drive the demand for air travel. We can only speculate at this point, and it remains to be seen what the long-term impact of the

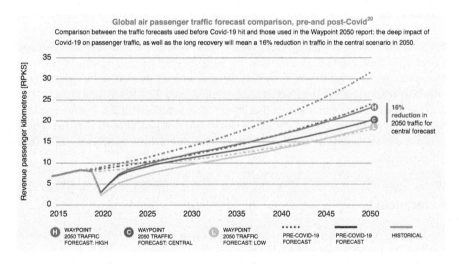

Fig. 9 Global air passenger traffic forecast comparison, pre- and post-COVID. Source: Air Transport Action Group (ATAG) (2020). Waypoint 2050. https://aviationbenefits.org/media/167187/w2050_full.pdf

COVID-19 pandemic will be on the aviation climate debate. However, the pandemic highlights the importance of all players in the aviation system take effective and efficient measures to reduce the negative climate impact of aviation. Our book aims to make a contribution in this regard.

6   Towards Sustainable Aviation

The rapid and profound changes can be overwhelming for stakeholders in the aviation system. In particular, the great interdependencies between the individual subsectors make it difficult to maintain an overview of developments. In addition, there is the danger of silo thinking, where exponents from politics, science, society, and industry deal with solutions for this highly relevant topic in parallel without coordinating with each other. Such an approach runs the risk of leading to sub-optimal results and provides the motivation for this book. The aim of this book is to analyze the topic "Sustainable Aviation" from a systemic management perspective. Our goal is to provide an overview of the most important topics and to carve out the central problems from a social science viewpoint, rather than a natural science or engineering viewpoint. In doing so, we also aim to outline possible pathways towards sustainable aviation including concrete courses of action.

The systemic approach taken in this book follows the general logic of the St. Gallen Management Model (SGMM) which views value creation for a dynamic environment as a complex interaction of different value creation processes. It considers not only the organization and its management but also the environment and their interaction. This systemic logic makes it clear that sustainability in aviation can only be achieved through a collaborative and coordinated approach among all stakeholders.

The book is roughly divided into two parts. In Part 1, we provide an overview of the technological developments for sustainable aviation. We then take the customer perspective and explore the question of whether and how customer behavior is changing. We contrast the two perspectives and address the debate on the extent to which technological progress needs to be complemented by behavioral change. Part 2 of the book is then devoted specifically to the stakeholders in the aviation system. We take an in-depth look at the strategies of airlines and airports, followed by a comprehensive analysis of the political and regulatory processes and their effectiveness.

References

Air Transport Action Group [ATAG]. (2020). *The economic and social benefits of air transport 2019*. Air Transport Action Group (ATAG).

Airbus. (2019). *Global market forecast*. Cities, Airports & Aircraft 2019-2038. Retrieved from https://www.airbus.com/aircraft/market/global-market-forecast.html

Cairns, S., & Newson, C. (2006). *Predict and decide: Aviation, climate change and UK policy*. Environmental Change Institute, University of Oxford.

Conrady, R., Fichert, F., & Sterzenbach, R. (2019). *Luftverkehr: Betriebswirtschaftliches Lehr-und Handbuch.* Walter de Gruyter GmbH & Co KG.

Cox, B. & Althaus, H.J. (2019). *How to include non-CO_2 climate change contributions of air travel at ETH Zurich.* https://doi.org/10.13140/RG.2.2.30222.10565

Elkington, J. (1997). *Cannibals with forks. The triple bottom line of 21st century business.* Capstone.

Elkington, J. (2018). 25 years ago I coined the phrase "triple bottom line." Here's why it's time to rethink it. *Harvard Business Review, 2*(5), 2–5.

Federal Office of Civil Aviation [FOCA]. (2021). *Luftfahrt und Klimaerwärmung.* Retrieved from https://www.bazl.admin.ch/bazl/de/home/politik/umwelt/luftfahrt-und-klimaerwaermung.html

Giddings, B., Hopwood, B., & O'Brien, G. (2002). Environment, economy and society: Fitting them together into sustainable development. *Sustainable Development, 10*(4), 187–196.

Gössling, S., & Humpe, A. (2020). The global scale, distribution and growth of aviation: Implications for climate change. *Global Environmental Change, 65*, 102194.

Graver, B., Zhang, K., & Rutherford, D. (2019). *CO_2 emissions from commercial aviation, 2018.* Retrieved from http://sustainablehighworth.co.uk/wp-content/uploads/2020/10/ICCT_CO2-commercl-aviation-2018_20190918.pdf

Hart, S. L. (1995). A natural-resource-based view of the firm. *Academy of Management Review, 20*(4), 986–1014.

Intergovernmental Panel on Climate Change. (1996). In J. K. Houghton, L. G. Meira Filho, B. A. Callander, N. Harris, A. Kattenberg, & K. Maskell (Eds.), *Climate change 1995. The science of climate change. Contribution of working group I to the second assessment report of the Intergovernmental Panel on Climate Change.* Cambridge: Cambridge University Press. Retrieved from https://www.ipcc.ch/site/assets/uploads/2018/02/ipcc_sar_wg_I_full_report.pdf

Intergovernmental Panel on Climate Change. (2013). Summary for policymakers. In T. F. Stocker, D. Qin, G.-K. Plattner, M. Tignor, S. K. Allen, J. Boschung, A. Naules, Y. Xia, V. Bex, & P. M. Midgley (Eds.), *Climate change 2013: The physical science basis. Contribution of working group I to the fifth assessment report of the intergovernmental panel on climate change* (pp. 3–29). Cambridge University Press. Retrieved from https://www.ipcc.ch/site/assets/uploads/2018/02/WG1AR5_SPM_FINAL.pdf

Intergovernmental Panel on Climate Change. (2014). In O. Edenhofer, R. Pichs-Madruga, Y. Sokona, E. Farahani, S. Kadner, K. Seyboth, et al. (Eds.), *Climate change 2014: Mitigation of climate change. Working group III contribution to the fifth assessment report of the Intergovernmental Panel on Climate Change.* Cambridge: Cambridge University Press. Retrieved from https://www.ipcc.ch/site/assets/uploads/2018/02/ipcc_wg3_ar5_full.pdf

Intergovernmental Panel on Climate Change. (2018). In V. Masson-Delmotte, P. Zhai, H.-O. Pörtner, D. Roberts, J. Skea, P.R. Shukla, A. Pirani, W. Moufouma-Okia, C. Péan, R. Pidcock, S. Connors, J.B.R. Matthews, Y. Chen, X. Zhou, M.I. Gomis, E. Lonnoy, T. Maycock, M. Tignor, & T. Waterfield (Eds.), *Global warming of 1.5°C. An IPCC special report on the impacts of global warming of 1.5°C above pre-industrial levels and related global greenhouse gas emission pathways, in the context of strengthening the global response to the threat of climate change, sustainable development, and efforts to eradicate poverty.* Retrieved from https://www.ipcc.ch/sr15/

Intergovernmental Panel on Climate Change. (2020). *Future pathways for adaptation, mitigation and sustainable development.* Retrieved from https://ar5-syr.ipcc.ch/topic_pathways.php

International Civil Aviation Organisation [ICAO]. (2016). *Resolution A40–19 CORSIA.* Retrieved from https://www.icao.int/environmental-protection/Documents/Assembly/Resolution_A40-19_CORSIA.pdf

International Civil Aviation Organisation [ICAO]. (2020). *World Aviation and the World Economy.* Retrieved from https://www.icao.int/sustainability/pages/facts-figures_worldeconomydata.aspx

International Civil Aviation Organisation [ICAO]. (2021). *Economic impacts of COVID-19 on civil aviation.* Retrieved from https://www.icao.int/sustainability/Pages/Economic-Impacts-of-COVID-19.aspx

International Energy Agency [IEA]. (2020). *Transport sector CO_2 emissions by mode in the sustainable development scenario, 2000–2030*. Paris: IEA. Retrieved from https://www.iea.org/data-and-statistics/charts/transport-sector-CO_2-emissions-by-mode-in-the-sustainable-development-scenario-2000-2030

Jungbluth, N., & Meili, C. (2019). Recommendations for calculation of the global warming potential of aviation including the radiative forcing index. *The International Journal of Life Cycle Assessment, 24*(3), 404–411. https://doi.org/10.1007/s11367-018-1556-3

Kurth, T., Wübbels, G., Portafaix, A., Meyer Zum Felde, A., & Zielcke, S. (2021). The *biodiversity crisis is a business crisis*. Retrieved from https://www.bcg.com/publications/2021/biodiversity-loss-business-implications-responses

Lovejoy, T. E. (2006). *Climate change and biodiversity*. TERI Press.

Morrison, S. A., & Winston, C. (1997). *The evolution of the airline industry*. The Brookings Institution.

Peet, J. (2009). *Strong sustainability for New Zealand: Principles and scenario. SANZ (Phase-2) report*. Nakedize Ltd.

Porter, M. E., & Kramer, M. R. (2006). The link between competitive advantage and corporate social responsibility. *Harvard Business Review, 84*(12), 78–92.

Rogelj, J., Den Elzen, M., Höhne, N., Fransen, T., Fekete, H., Winkler, H., & Meinshausen, M. (2016). Paris agreement climate proposals need a boost to keep warming well below 2 °C. *Nature, 534*(7609), 631–639. https://doi.org/10.1038/nature18307

Ryley, T., Baumeister, S., & Coulter, L. (2020). Climate change influences on aviation: A literature review. *Transport Policy, 92*, 55–64. https://doi.org/10.1016/j.tranpol.2020.04.010

Seinfeld, J. H., & Pandis, S. N. (2006). *Atmospheric chemistry and physics: From air pollution to climate change* (2nd ed.). Wiley.

Shepherd, B., Shingal, A., & Raj, A. (2016). *Value of air cargo: Air transport and global value chains*. The International Air Transport Association (IATA).

Sims, R., Schaeffer, R., Creutzig, F., Cruz-Núñez, X., D'Agosto, M., Dimitriu, D., Figueroa Meza, M. J., Fulton, L., Kobayashi, S., Lah, O., McKinnon, A., Newman, P., Ouyang, M., Schauer, J. J., Sperling, D., & Tiwari, G. (2014). Transport. In O. Edenhofer, R. Pichs-Madruga, Y. Sokona, E. Farahani, S. Kadner, K. Seyboth, A. Adler, I. Baum, S. Brunner, P. Eickemeier, B. Kriemann, J. Savolainen, S. Schlömer, C. V. Stechow, T. Zwickel, & J. C. Min (Eds.), *Climate change 2014: Mitigation of climate change. Contribution of Working Group III to the Fifth Assessment Report of the Intergovernmental Panel on Climate Change* (pp. 599–670). Cambridge University Press.

Stern, N. H. (Ed.). (2007). *The economics of climate change: The Stern review*. Cambridge University Press.

Umweltbundesamt. (2019). *Umweltschonender Luftverkehr. Lokal – national – international* [Report]. Retrieved from https://www.umweltbundesamt.de/sites/default/files/medien/1410/publikationen/2019-11-06_texte-130-2019_umweltschonender_luftverkehr_0.pdf

UNFCCC. (2015). Adoption of the Paris Agreement. Report No. FCCC/CP/2015/L.9/Rev.1. Retrieved from https://unfccc.int/resource/docs/2015/cop21/eng/l09r01.pdf

United Nations. (1992). *United Nations Framework Convention on Climate Change* [Convention protocol]. Retrieved from https://unfccc.int/resource/docs/convkp/conveng.pdf

United Nations. (1998). *Kyoto protocol to the United Nations framework convention on climate change*. Retrieved from https://unfccc.int/resource/docs/convkp/kpeng.pdf

Wittmer, A., Bieger, T., & Mueller, R. (2021). *Aviation systems – Management of the integrated aviation value chain*. Springer.

World Business Council for Sustainable Development, & World Resources Institute. (Eds.). (2004). *The greenhouse gas protocol: A corporate accounting and reporting standard* (Revised edn). : World Resources Institute.

World Commission on Environment and Development. (1987). *Our common future*. Oxford University Press.

Technology Assessment for Sustainable Aviation

Alexander Stauch and Adrian Müller

Abstract

- The most promising technology to increase sustainability in aviation in the short term is sustainable aviation fuel (SAF).
- In the long run, battery technology could be used for short-haul flights while hydrogen-based aircraft could serve long-distance flights.
- A competition between hybrid (SAF and battery) and hydrogen technology is likely to happen in the mid-run for medium and long-distance flights.
- The long-term development of alternative technologies remains in general uncertain and is highly dependent on the investments of large industry incumbents (e.g., Airbus and Boeing), new start-ups but also on the development of other industries (regarding hydrogen and battery).
- Therefore, investment executives are well-advised to allocate resources and build connections to other industries to further explore the sustainability and economic potential of new technologies (particularly hydrogen and all-electric aircraft).

Since almost all GHG emissions in the aviation sector are attributable to the usage of fossil jet fuel, increased efficiency measures, as well as alternatives to conventional jet fuel, become highly relevant to reduce GHG emissions from aviation, in order to reach global climate goals. Consequently, this chapter analyses different opportunities based on technology to reduce the overall consumption of fossil jet fuel. These technologies mainly include aircraft propulsion

A. Stauch (✉)
Institute for Economy and the Environment, University of St. Gallen, St. Gallen, Switzerland
e-mail: alexander.stauch@unisg.ch

A. Müller
Center for Aviation Competence, University of St. Gallen, St. Gallen, Switzerland
e-mail: adrian.mueller@unisg.ch

systems, alternative fuels, aircraft models based on alternative energy usage, new aircraft design (blended-wing body aircraft), and supportive technologies for efficient ground handling and optimized operations during flight. Thus, the focus of this chapter lies on aircraft technologies and technologies related to flight operations.

1 Overview of Current and Future Propulsion Technologies for Sustainable Aviation

This chapter investigates the different aircraft technologies and their primary source of energy. We start with a general overview on technology for propulsion, continuing with outlining alternative fuels to conventional jet fuel, followed by an outline of new aircraft models based on alternative energy sources such as electricity and hydrogen. The last subchapter concludes this section with an overview of alternative aircraft model designs.

1.1 Aircraft Propulsion Technologies

To understand how aviation can be made more sustainable in the future, it is fundamentally important to be familiar with the different aircraft propulsion systems and their required energy sources. Similar to the automotive sector, propulsion systems for commercial aviation applications can be divided into internal combustion engine, hybrid-, and all-electric propulsion systems. Unsurprisingly, the internal combustion engine is the dominant form of propulsion system for commercial aircraft today.

1.1.1 Conventional Internal Combustion Engines
The conventional internal combustion engine, as the name suggests, generates energy by burning fuel. In aviation, the fuel is in most cases kerosene, also referred to as jet fuel. In general, an internal combustion engine can be powered by all sorts of burnable and liquid fuels, including oil, gasoline, diesel, kerosene but also biofuels, synthetic fuels, and hydrogen.

Manufacturers are still striving to improve the efficiency of combustion engines. However, potential efficiency gains with conventional internal combustion engines are already very limited. Since major efficiency gains have already been achieved, today's focus lies mainly on the reduction of fuel and kerosene consumption per thrust at a certain speed. The main advantage of an internal combustion engine is the high energy density of the required energy source used to burn, the ranges that can be achieved with it, and its cost-efficient and global availability. However, the rather low level of energy efficiency (between 35 and 45 percent), a comparably high noise level and the emission of greenhouse gas emissions resulting from burning fuel represent the main disadvantages.

1.1.2 All Electric Propulsion Systems

The operating principle of electrified propulsion systems relies on the production, storage, and transmission of electrical power (electricity). Therefore, fully electrified aircraft propulsion systems are completely based on electric motor engines instead of internal combustion engines. The required energy for the propulsion systems is electricity, which is stored in one or several batteries. Similar to refueling, batteries can be recharged. Electricity can have different kinds of origins, ranging from renewable sources that include wind, solar, and hydro to fossil-based electricity from burning coal, oil, or gas.

The biggest advantage of electric motor engines is the high level of energy efficiency (up to 95%) while the biggest disadvantage is represented by the low energy density of the battery that is required to power the engine, which results in lower operating times and ranges due to higher weight. Further notable advantages of electric propulsion systems are lower maintenance cost, lower noise levels, and of course almost no greenhouse gas emissions (depending on the source of electricity). An additional disadvantage is the nonexistent recharging infrastructure.

1.1.3 Hybrid Propulsion Systems

While today's commercial aircraft are in most cases only equipped with an internal combustion engine, the hybrid form of propulsion uses a combustion engine in combination with an electric propulsion system.

The main advantages of hybrid propulsion systems are on the one hand significantly lower fuel consumption compared to purely combustion-powered aircraft and on the other hand lower noise pollution due to quieter electric propulsion. Maintenance costs are also likely to decrease, as electric propulsion systems require less maintenance efforts. Thus, maintenance efforts are also dependent on the amount of use of the combustion engine. The disadvantages cover the fact that both, electric and conventional propulsion components must be duplicated to some extent, which increases weight, for example, and therefore results in rather limited ranges, mainly applicable for short-to-medium distances.

1.2 Sustainable Aviation Fuels

We define Sustainable Aviation Fuel (SAF) as a "drop-in" fuel that meets all the same technical and safety requirements as fossil-based jet fuel. SAF contains the same hydrocarbons (and thus the same tailpipe emissions) as fossil-based kerosene, but the difference is that the hydrocarbons came from a more sustainable source, such as biological organisms, water, and air. This can result in a net reduction of emissions up to 80% compared to fossil jet fuel on a life cycle basis.

Alternative fuels aim to reduce greenhouse gas emissions and are characterized by the use of renewable resources; they should be fully compatible with the technology available today, namely with the internal combustion engine, and are favorable for the reduction of exhaust pollutant emissions. Classification of alternative fuels compared to fossil fuels can be made into biofuels and synthetic fuels. Both types of fuel feature "drop-in characteristics," which means that the fuels can be used in exactly the same manner as, and even be blended with, regular aviation fuel with the technology and infrastructure already available today. Compared to regular aviation fuel, synthetic and biofuel reduce greenhouse gas emissions significantly but are also much more expensive (Heyne et al., 2021).

1.2.1 Biofuels

The general idea of biofuel is simple and mainly environmentally motivated. Plants absorb carbon dioxide as they grow, enabling plant-based biofuels that emit the same amount of greenhouse gases as previously absorbed. Biofuel production, processing, and transport also emit greenhouse gases, reducing the emissions savings. During the biofuel production process, biomass is converted through advanced processes, covering both thermochemical and biochemical routes. Biofuels are technologically feasible, and certain types of biofuels are already certified and can be mixed with kerosene up to 50%. This certification refers to the year 2019, but it is assumed that this might change to higher values in the near future since a successful commercial flight with 100% biofuel usage was already accomplished by Boeing and FedEx in 2018. Boeing, for instance, announced new aircraft models capable of 100% biofuel usage by the end of this decade.

A distinction is made between three generations of biofuels. The first generation involves the conversion of plants into biofuel, which can also serve nutritional purposes. This may give rise to sustainability concerns in connection with competition with food production. The second generation involves the conversion of recycled oils (used cooking oil), residual materials (municipal solid waste), or nonedible plants. Feedstocks for this generation of biofuel are lignocellulosic biomass (dedicated or residual) and other more unconventional raw materials such as algae. These feedstocks do not compete with sugar or other oil-based plants. However, the planting of crops for biofuels can have negative effects such as deforestation and soil erosion, as well as increasing pressure on water resources through agriculture and refineries.

The third and most recent generation of biofuels includes new types of biofuels made, for example, from algae or special oil plants. They are not cultivated in a field, but in a bioreactor or on fallow land, for instance, in the desert. Flexibility concerning the location of the production site is intended to reduce the high costs of transporting biomass on a large scale and thereby contribute to economies of scale for biofuel. Biofuels with the highest emission savings are those derived from photosynthetic algae (98% savings) and from nonfood crops and forest residues (91–95% savings). However, the production of SAF generates emissions that are caused by the equipment required to grow the plants, transport the raw materials, refine the fuel, etc. If these elements are considered, it has been shown that the use of sustainable aviation fuel still leads to a significant reduction in the total lifecycle of net[1] CO_2 emissions compared to fossil fuels, which in some cases can be as much as 80%.

1.2.2 Synthetic Fuels

Synthetic fuel is different from biofuel with respect to its ingredients, as synthetic fuel is not based on biological organisms (such as feedstocks or algae). Synthetic fuel is purely based on water (H_2O) and CO_2 from the surrounding air, and on the use of high temperatures for its production. The primary energy used to produce synthetic fuels allows this category to be further divided into solar and electric fuels. Depending on the primary source, whether solar energy or electricity is used, a distinction is made between the following manufacturing processes: sun-to-liquid (STL) or power-to-liquid (PTL). Both production processes are based on the idea of converting water (H_2O) and carbon dioxide (CO_2) under very high temperatures into hydrogen (H_2), carbon monoxide (CO), and Syngas, a mixture of mainly H_2 and CO which is subsequently processed to so-called Fischer-Tropsch hydrocarbon fuels. The supply of carbon dioxide (CO_2) for the synthesis process can potentially be achieved by CO_2 extraction from the air. The underlying principle behind the manufacturing path is to absorb CO_2 from the atmosphere during manufacturing and to emit only the same quantity of CO_2 during combustion, thus achieving a neutral CO_2 cycle. Moreover, synthetic fuels are cleaner than fossil fuels. They contain less heavy metal and sulfur contaminants than petroleum fuels and thus burn cleaner with less residue such as sooth or sulfur dioxide. Synthetic aviation fuel is compatible with todays internal combustion engines and does therefore not require new aircraft and propulsion technologies. While the sun-to-liquid process requires fewer production steps and offers therefore greater efficiency compared to power-to-liquid, both production pathways lead to the same outcome: synthetic aviation fuel based on the production of syngas. Solar conversion efficiency and the cost of a sustainable CO_2 supply are key factors in the competition with fossil and biomass-based fuels.

1.3 New Aircraft Technologies and Alternative Energy Repository

Besides alternative and sustainable aviation fuels, there are also other sources of energy that are suitable to effectively power propulsion systems, either with zero CO_2 emissions (hydrogen) or without any emissions at all (electricity). From an emissions perspective, both energy sources offer superior advantages over kerosene, biofuel, and synthetic fuel. The major disadvantage for both energy sources is the requirement of new aircraft models and heavy investments into the airport and energy infrastructure.

---

[1] Net emissions mean that CO_2 emissions from the natural CO_2 cycle are treated as net zero emissions. Thus, a net emission reduction is not the same as an absolute emission reduction of emissions.

1.3.1 Electricity and Battery

Powering an all-electric aircraft with batteries only, using electricity generated from renewable sources, belongs to the so-called "true zero solutions," meaning that all gross emissions are reduced to zero during operation. The concept is based on electricity stored in a battery that powers an electric engine for propulsion. Despite all noteworthy advantages of electric motor engines, such as higher efficiency, less noise, lower maintenance, and the use of renewable power sources to achieve zero gross emissions, the big drawback is the low energy density (or power-to-weight ratio) of the battery.

For a use of an all-electric propulsion system in larger commercial aircraft, the technical demand regarding currently available battery technology, mainly lithium-ion battery systems, is too great. A high power-to-weight ratio is crucial for the practical use of battery systems in commercial aviation, but the lithium-based technology only achieves 150–250 Wh/kg. In contrast, the current state of technology would require a minimum of 800 Wh/kg for hybrid systems and 1800 Wh/kg and more for all-electric systems. Or from another perspective: the Airbus A320 has a maximum take-off mass of 78 tons, while the weight of the battery to power an Airbus A320 fully and only with electricity would be around 287 (based on tomorrow's projected best batteries) and 1'293 tons (based on today's Tesla Model S batteries). The current state of technology only allows small all-electric planes, with a maximum of 6 passenger seats. In 2019, for example, a short test flight by Harbour Air and magniX involved a six-passenger aircraft fitted with an electric motor. Further small and fully electric plane models developed by different companies are in the pipeline (e.g., easyJet and Wright Electric).

> **Wright Electric and easyJet Case**
>
> When it comes to all-electric commercial aircraft seating up to 200 passengers, easyJet, in collaboration with the Californian start-up Wright Electric, already announced its first joint technology development projects in 2018.
>
> As of 2021, the electric aircraft start-up Wright Electric has started to develop the electric propulsion system for a 186-seat electric aircraft called "Wright 1," which should cover short-haul routes by 2030 (Fig. 1). Specifically, this involves a 1.5-MW electric motor and a 3-kilovolt inverter. This aircraft will be able to cover distances of up to 600 kilometers with one charging cycle. CEO Johan Lundgren is already very excited: "The target range of the electric plane is around 500 km, which, within our current route portfolio, would mean a route like Amsterdam to London could become the first electric 'flyway'."
>
> Regarding the battery, easyJet and Wright Electric are confident that the required energy density for the battery will be achieved until 2030. "Battery technology is advancing at pace with numerous US government agencies now funding research into electric aviation—all of these developments help us to more clearly see a future of more sustainable operations," said easyJet CEO Johan Lundgren.

Fig. 1 The "Wright 1" Model, a 186-seat all-electric aircraft. Source: Business Traveller (2020)

1.3.2 Hydrogen for Combustion

Another approach to replacing kerosene in the aviation industry would be to switch to complete CO_2-neutral energy carriers, such as hydrogen. A hydrogen-powered aircraft is not a new technology but has been tested already: a Soviet Tupolev made the first hydrogen-fueled flight in 1988. Since this happened during the time of the Soviet Union, concrete data about the flight performance remained sealed by the Soviet Government. With the end of the Soviet Union, however, silence fell on the Tupolev and its planned successor models.

Hydrogen, in its liquid form LH_2, could be used in the same way as conventional fuels in a combustion engine and would only exhaust Water (H_2O) and Nitrogen Oxide (NO_x) but zero CO_2. Water vapor (around 9 kg for 1 kg of combusted hydrogen) and NO_x still contribute to greenhouse gas levels in our atmosphere and are, similar as regular fuel emissions, also producing contrails and aviation-induced cloudiness. The extent of aviation-induced cloudiness caused by the combustion of hydrogen compared to fossil jet fuel is not yet fully investigated, but researchers argue that it should be less harmful than aviation-induced cloudiness from fossil jet fuel, as hydrogen has no soot particles. Additionally, there are some technologies expected that could store the water vapor from operations, meaning the effect on aviation-induced cloudiness will be drastically decreased to fossil jet fuel. Further, due to its low volumetric density (Fig. 2), liquid hydrogen requires four to five times the volume of conventional fuel to carry the same onboard energy and needs to be kept at $-253\ °C$ to maintain its liquid physical state.

The advantage of liquid hydrogen is its weight: it is approximately 2.8 times lighter than conventional kerosene. This could partially equalize the disadvantages of its lower energy density. After all, liquid hydrogen remains superior to

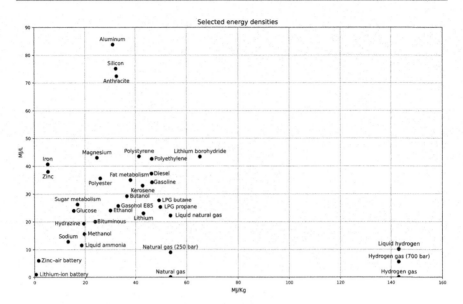

Fig. 2 Energy density based on power-to-weight ratios of different energy sources. Source: Dial (2011)

conventional fuel in terms of power density by unit weight. This is highly relevant for flight, a weight-critical application, as it offers a Maximum Take-off Weight (MTOW) advantage over all other energy storage alternatives.

The main barrier of an industry-wide hydrogen use is the fact that liquid hydrogen needs to be stored at a low temperature and in cylindric pressure containers, which leads to new challenges when it comes to requirements for isolation and containment systems as well as the engine parts that need to operate under extreme thermal conditions. In terms of the aircraft's structural design, significant adjustments are required to support the use of LH_2 tanks. The wing space currently used to contain fuel will not be suitable for the large and heavy cylindric containers that are necessary to store liquid hydrogen. Aircraft models will thus have to change along with that technology.

1.3.3 Hydrogen for Fuel Cell

A fuel cell converts hydrogen and oxygen (from air) into electricity. The electricity then powers an electric motor engine to run the aircraft. This potentially offers a zero-emission technology, since the only "emission" of the fuel cell is water. However, the water produced, around 9 kg for every 1 kg of hydrogen reacted (same as hydrogen for combustion), would have to be released, and water vapor is also a GHG with the additional potential to cause contrails and aviation-induced cloudiness.

Experts also believe that hydrogen fuel cell aircraft would be even more efficient than hydrogen combustion designs, needing to carry 20–40 percent less fuel (hydrogen). On the other hand, the use of hydrogen and fuel cell technology for aviation requires an even more drastic aircraft redesign, compared to hydrogen for combustion, to accommodate the electric and distributed propulsion system, the full suite of new electric sub-systems and the hydrogen storage. Further, the development of aviation-ready, efficient, power-dense fuel cells as not yet reached satisfactory results. Since hydrogen for combustion does not require as drastic a change in infrastructure and aircraft technology, is less efficient and not as clean as hydrogen for fuel cell, the future development of both technologies will depend on the views and priorities set by incumbent and new aircraft manufactures. At the moment, large manufacturers (e.g., Airbus) develop hydrogen for combustion aircraft as this is believed to be sooner ready for market, while smaller companies or new companies also started to develop fuel cell-based aircraft (e.g., ZeroAvia). The bottom line, no matter whether the fuel cell takes off or not, is that hydrogen currently offers the greatest potential for reducing greenhouse gases in aviation in the long run.

Airbus Case

Toulouse, September 21, 2020—Airbus has revealed three concepts for the world's first zero-emission commercial aircraft which could enter service by 2035. These concepts each represent a different approach to achieving zero-emission flight, exploring various technology pathways and aerodynamic configurations in order to support the company's ambition of leading the way in the decarbonization of the entire aviation industry.

All of these concepts rely on hydrogen for combustion as a primary power source—an option which Airbus believes holds exceptional promise as a clean aviation fuel and is likely to be a solution for aerospace and many other industries to meet their climate-neutral targets.

"This is a historic moment for the commercial aviation sector as a whole and we intend to play a leading role in the most important transition this industry has ever seen. The concepts we unveil today offer the world a glimpse of our ambition to drive a bold vision for the future of zero-emission flight," said Guillaume Faury, Airbus CEO. "I strongly believe that the use of hydrogen—both in synthetic fuels and as a primary power source for commercial aircraft—has the potential to significantly reduce aviation's climate impact."

Airbus announced three different new aircraft concepts all codenamed "ZEROe." They can carry 100–200 passengers with ranges between 1000 and 2000 nautical miles (1800–3600 km) (Fig. 3).

1.4 Alternative Aircraft Concepts: Blended-Wing Body Aircraft

The blended-wing body concept is associated with an aircraft consisting of a single wing. This design represents a highly coupled system in which the functional elements for thrust, control, and lift are integrated into the wing. The integration of

Fig. 3 Airbus ZEROe Hydrogen Aircraft Designs. Source: Airbus Company Homepage (2021)

all sub-systems, such as engines, passenger cabin, cargo compartment, and control surfaces, into the wing allows for a minimization of drag and an improvement in aerodynamic efficiency. The latter can be increased by up to 15% with this concept compared to a conventional tube and wing architecture. At the same time, the take-off weight could be reduced by up to 10%. These optimizations could reduce fuel consumption and thus CO_2 and NO_X emissions.

The design also enables new forms of propulsion integration, such as distributed propulsion or boundary layer ingestion,[2] which in turn could have a positive effect on noise emissions. Depending on the position of the engines, there are advantages and disadvantages concerning accessibility, noise emissions, and safety-related matters. Due to improved aerodynamic efficiency (better lift-to-drag ratio) compared to a conventional tube and wing architecture (TAW), better operational efficiency can be achieved. Recent studies assume a reduction in fuel consumption of 20–50%

---

[2] On today's jet aircraft, the engines are typically located away from the aircraft's body to avoid ingesting the layer of slower flowing air that develops along the aircraft's surfaces, called the boundary layer. The new propulsor design (boundary layer ingestion), meaning the inlet and the fan, is embedded in the aircraft body at the back of the fuselage and ingests the slower boundary layer air flow, using it to generate the thrust needed to propel the aircraft through its mission. Using the slower boundary layer air means the engines do not have to work as hard, so their fuel consumption goes down. At the same time, the drag on the aircraft itself is reduced, since the engines "ingested" part of that drag, so the overall fuel efficiency of the aircraft is better and less thrust is needed by the aircraft to fly at the same speed. This means the overall aircraft efficiency is higher and less fuel is needed to complete the flight (NASA, 2021).

Fig. 4 Hydrogen-powered blended-wing body design by Airbus. Source: Airbus Company Homepage (2021)

compared to the classic tube and wing design. A study by Zhenli et al. (2019) assumes an average fuel saving of 31.5% compared to a TAW design. However, a lack of experience and empirical data exists in connection with the poor stability inherent in the concept. The certification faces various technological challenges, such as the evacuation of passengers in an emergency or the development of a pressurized cabin that does not have an almost circular cross section as in TAW configurations. Another key sticking point is customer acceptance, as passengers sitting on the outside would likely be exposed to greater G-forces and a more unsteady flight experience.

The blended-wing body aircraft could be powered with regular kerosene, bio-, and synthetic fuels but also with hydrogen- and battery-powered electricity. Since the blended-wing body concept offers greater energy efficiency, the concept would be suitable for long-distance flights as it can fly further with the same amount of energy as conventional tube and wing concepts. This is particularly interesting for battery and hydrogen applications, as both of them require more onboard space than aviation fuel. This might also be the reason, why Airbus announced a blended-wing body aircraft based on hydrogen to cover long-distance flights (more than 2000 nautical miles (over 3600 km)) (Fig. 4).

2 Technology Evaluation

In this subchapter, the introduced aviation technologies from the previous subchapter will be evaluated and compared with respect to sustainability impact (GHG emission reduction), cost-competitiveness and its long-term development projections as well as time to market with respect to sustainability impact. This subchapter will be closed by a general conclusion on current and future aircraft technologies as well as on alternative energy usage.

2.1 Sustainability Evaluation

In this section, emission reductions based on the natural CO_2 cycle are labeled as "net reduction," while a real reduction based on emissions prevention is labeled as "absolute reduction" in GHG emissions.

Kerosene is a very energy-dense fuel particularly for aviation based on oil, which is a fossil energy source. Due to the very good combustion of kerosene, the exhaust gases consist of almost 100 percent hot air. This includes the nontoxic gases carbon dioxide (CO_2; 7 percent) and water vapor (3 percent), which have a direct impact on the climate and are produced during the combustion of hydrocarbons from kerosene. The actual toxic pollutants account for only about 0.04 percent of the exhaust gases and include NO_X, CO, sulfur dioxide, unburned hydrocarbons, and particulate matter (soot particles).

Biofuels with the most net emission savings are those derived from photosynthetic algae (98% savings) and from nonfood crops and forest residues (91–95% savings). But there is also the need to take a holistic view of the supply chain into account for assessing the sustainability of biofuel, which includes land-water usage, GHGs and particle emissions, as well as competition with agriculture (fuel-food, including animal food, competition). By blending biofuels up to 50% with conventional fuel, the ecological net footprint of the blended fuel can be reduced by up to 30% to 40% over its life cycle, depending on the production method and feedstock used. Assuming a 100% usage of biofuel during flight, it could even reduce the net GHG emission up to 80% from a total life cycle perspective.

Synthetic Fuel based on a sun-to-liquid production has the advantage of taking the needed resources for production directly from the surrounding air and sunlight, and therefore does not need any limited or environmentally problematic input factors. Due to this production process, synthetic fuel has a CO_2 neutral footprint, as it uses the same amount of CO_2 that is emitted during combustion for its production. Overall synthetic fuel also claims to reduce net GHG emissions up to 50% in general compared to conventional kerosene, while the remaining CO_2 emissions from synthetic fuel are part of the natural CO_2 cycle. However, it must be also noted that synthetic fuel still produces other pollutants during combustion that are not part of a natural resource cycle (water vapor, NO_X, CO, soot particles), but this share is still 50% lower than for conventional fossil jet fuel. In total, synthetic fuel can reduce net GHG emissions up to 80% compared to conventional kerosene.

However, as there are still high net emissions during the flight, the use of bio- and synthetic fuels still contributes to climate change, as these emissions promote contrails and aviation-induced cloudiness, which in turn have a strong influence on radiative forcing (change in the earth's energy balance due to a change in the effect of radiation from space caused by, e.g., aviation-induced cloudiness from fuel emissions). If radiative forcing increases, the earth's temperature will rise and thus further contribute to the greenhouse effect.

Hydrogen itself is completely climate neutral since it consists only of water and oxygen. Burning hydrogen does not emit CO_2 at all. The only emissions from the combustion of hydrogen are NO_x (due to micro-mix combustors, lower levels of NO_x than modern kerosene engines could be produced) and water vapor, which both contribute to the greenhouse gas effect by causing contrails and aviation-induced cloudiness (reinforced radiative forcing). Water vapor, with the potential to cause contrails and aviation-induced cloudiness, is also a greenhouse gas, however, its harmful effects can be minimized through careful operation. When hydrogen is converted into electricity in a fuel cell, the only remaining emission is water vapor. A hydrogen fuel cell aircraft could be designed to store some of the water produced and release it in conditions conducive to low contrail/aviation-induced cloudiness formation. This concept has not yet been proven in practice and therefore needs further research to be fully understood. In general, radiative forcing is assumed to be significantly lower than for other fuel types, including bio- and synthetic fuels, as emissions from burning hydrogen do not contain any soot particles, CO_2, CO, sulfur dioxide, and unburned hydrocarbons.

Electricity and Battery offer the greatest potential with regard to sustainability, meaning it is the only technology that can reduce all emissions during flying to zero. The use of electricity, which comes from a battery, in an electric motor engine does not cause any emissions. Therefore, the use of electricity from renewable sources, such as solar, wind, or hydrogen, is crucial to sustainability and climate neutrality. Of course, one could now argue that the production of renewable energy generators is not CO_2 neutral. However, this depends on the producers. For example, a solar cell producer can decide for themselves whether they produce their solar cells with renewable or other energies. The production of solar power by the solar cells always remains CO_2 neutral and does not produce any other emissions.

Hybrid aircraft models, based on the usage of 50% synthetic aviation fuel on the one side and 50% power usage from battery-powered electricity on the other side, offer an interesting combination of benefits regarding sustainability and technological feasibility. Hybrid planes powered by combusting synthetic aviation fuel during energy intense periods of the flight (take-off and landing) could reduce GHG emissions in critical heights, since the emissions coming from the electricity used to power the electric engine during the main part of the flight would be zero. In comparison with 100% kerosene, such a hybrid configuration could reduce net GHG emissions by around 90% (if CO_2 emissions from the natural CO_2 cycle are treated as net zero emissions), and up to 75% in total. Regarding technological feasibility, hybrid planes do require a much smaller battery than an all-electric plane of the same size, which significantly lowers the take-off weight of the aircraft. However, there are still large improvements in battery technology required to achieve the technological feasibility of large hybrid commercial aircraft (Table 1).

Table 1 Sustainability evaluation of aviation power and fuel technologies

	CO_2 emissions per kWh of energy	Other relevant greenhouse gas emissions	Total effect on greenhouse gas effect during flying
Kerosene	Approximately 250 g of CO_2 per kWh	Water vapor, NO_X, CO, sulfur dioxide, unburned hydrocarbons, and particulate matter (soot particles)	Approximately 2.55 kg of CO_2 for 1 liter of kerosene, and other GHG emissions during flight and radiative forcing
Biofuel (50% mixed with kerosene)	Approximately 190 g of CO_2 per kWh, but 50% attributable to natural CO_2 cycle	Water vapor, NO_X, CO, sulfur dioxide, unburned hydrocarbons, and particulate matter (soot particles)	Up to 40% less net GHG emissions than kerosene and still radiative forcing
Biofuel (100%)	Net zero: Approximately 130 g of CO_2 per kWh, but part of natural CO_2 cycle	Water vapor, NO_X, CO, sulfur dioxide, unburned hydrocarbons, and particulate matter (soot particles)	Up to 80% less net GHG emissions than kerosene but still radiative forcing
Synthetic fuel (100%, based on solar power)	Net zero: Approximately 125 g of CO_2 per kWh, but part of natural CO_2 cycle	Water vapor, NO_X, CO, sulfur dioxide, unburned hydrocarbons, and particulate matter (soot particles)	Up to 80% less net GHG emissions than kerosene, but still radiative forcing
Hydrogen (produced with renewable energy)	Absolute zero	NO_X, water vapor Fuel cell: Only water vapor	Net-zero GHG emissions and absolute zero CO_2 but still radiative forcing (lower than for kerosene and SAF)
Electricity (solar)	Absolute zero	No other emissions	Absolute zero emissions during flight No radiative forcing
Hybrid: SAF plus electricity and battery (50%)	Net zero: Approximately 125 g of CO_2 per kWh from synthetic fuel, CO_2 neutral cycle	Water vapor, NO_X, CO, sulfur dioxide, unburned hydrocarbons, and particulate matter (soot particles)	95% less net GHG emissions than kerosene, if battery covers 50% of power usage but still radiative forcing

Sources: Lee et al. (2021), Liu et al. (2018), Magone et al. (2021), Pavlenko and Searle (2021), Quaschning (2021), Thomson et al. (2020)

2.2 Economic Evaluation

The economic evaluation takes two different perspectives on the cost. Energy production cost on the one side and investment costs in new infrastructure as well as in new aircraft technologies on the other side will be considered. It is important to consider both costs to get a full picture of long-term transformation costs and its required investments. The cost for energy production is given in USD per kWh, to ensure comparability among different energy sources. One liter of kerosene, for instance, contains approximately 10.35 kWh of energy.

The investment cost for new infrastructure covers fuel delivery to airports and airport refueling, while the investment cost for new aircraft technology reflects (1) the cost of buying new aircraft models for airlines that are capable with the new and alternative energy sources, such as hydrogen or electricity coming from a battery and (2), the research and development costs from plane manufacturers.

Kerosene is already at a very low-cost level. Depending on domestic taxes and regulations, the price for kerosene varies strongly between countries (0.03 to 0.07 USD per kWh of kerosene). It will not be very likely to see further price decreases of kerosene in the future, in contrast, when greater emissions sanctions are imposed on aviation, such as ETS (Emission Trading System) and CORSIA (Carbon Offsetting and Reduction Scheme for International Aviation), the operating cost of burning jet fuel is likely to rise. Still depending on domestic regulations, the prices of kerosene could go up between 40% and 100% until 2035, depending on national and international policy measures for CO_2 pricing. Such an increase in kerosene pricing would of course influence the competitiveness of alternative fuels and technologies in a positive way. However, this influence is rather difficult to predict, as we do not know exactly when and to which extent carbon pricing on aviation jet fuel will happen.

Biofuel starts at 0.07 USD per kWh and goes up to 0.16 USD/kWh. For advanced biofuels, the price difference is even more significant, starting at USD 0.1 USD/kWh going up to 0.25 USD/kWh. Since biofuels for aviation are at an early development stage, the future prices will be highly dependent on further investments into biofuel production. If scalability to the mass market is achieved, prices of biofuel will be competitive to the ones of kerosene. To achieve this, however, it still requires high investments in the infrastructure for production. At the moment, governments around the world are increasing their support for the production of alternative fuels and companies like Neste (2021) at the same time are making great progress on development processes for alternative fuels. Thus, a price decline in the near future is very likely to happen.

Synthetic fuel is currently far from economic feasibility, since the development is still at a very early stage. If economies of scale and efficiency gains are realized in the next years, synthetic fuel could reach kerosene's current price level by approximately 2035. The long-term development is highly dependent on future investor decisions, since the further development of synthetic fuel still requires large investments into the production infrastructure. On the other side, synthetic and also biofuel do not require new aircraft models, since both of them can be used with already existing combustion engines.

Hydrogen production, when produced from renewable sources, is currently rated between 0.10 and 0.14 USD per kWh. Similar to electricity and battery, hydrogen is likely to penetrate into other industries too, which could speed up the development of fuel cells and storage systems, promote downstream infrastructure and push down production costs. This would also benefit the aviation industry, as the R&D and infrastructure development costs would be partially borne by other industries. While demand for hydrogen from other transportation sectors increases, and supply rises in line with renewable energy capacity, costs will likely fall. For example, projects are under development in Australia, Saudi Arabia, and North Africa where "green" hydrogen is expected to cost as little as 0.07 USD/kWh in the near future. Additionally, technology improvements in electrolyzers and hydrogen compression methods as well as declining electricity prices for renewables (particularly solar electricity, which is often used to produce hydrogen) are also likely to contribute further to cost reduction. In 2035, hydrogen will be at the same price level or even cheaper than kerosene today.

Hydrogen production is often criticized for requiring too many power conversion steps, each of which diminish its overall production efficiency (and increase cost). For example, converting cheap solar electricity into expensive hydrogen may seem like a redundant and unprofitable step, to simply convert it back into electricity in a fuel cell. By contrast, employing a battery to power an aircraft would seem simpler and more efficient. However, if battery improvements plateau at a point insufficient for mid- to long-haul flight, hydrogen may remain the only "zero carbon" or "true zero emission" (we also refer to this as absolute zero) option.

One major barrier for a wide application of hydrogen in aviation, besides the required and high infrastructure investment for hydrogen production, distribution and storage (at airports and outside airports), is the need of new aircraft models, which are capable of dealing with the requirements of hydrogen as aviation fuel. This requires on the one hand large research and development costs by plane manufacturers, but on the other hand also a high investment by airlines to renew their fleets. Further, the cost and complexity increase drastically for fuel cell planes, sine fuel cell planes require the development of aviation-ready, efficient, power-dense fuel cells as well as improved electric motors, power electronics, cabling, and other electrical components. This ultimately requires an entirely new aircraft design that leverages distributed electrified propulsion to get all benefits of a fuel cell-powered aircraft.

Electricity already belongs to the cheapest sources of energy. Depending on the renewable energy source, production prices can range from 0.05 USD per kWh (optimal solar conditions) to 0.20 USD (weak solar conditions). It is assumed that the cost of solar power will, based on efficiency gains and economies of scale, further decrease, reaching 0.02 USD per kWh in 2035. Weak solar conditions usually offer better wind conditions on the other hand, leading to production prices of 0.06 USD per kWh (off-shore wind) and 0.10 USD per kWh (onshore wind). Hydropower prices usually depend on the size and capacity of the power plant and can range from 0.05 USD to 0.10 USD per kWh of produced energy. Similar as for hydrogen, the aviation industry could benefit from the developments and electrification of other industries, such as the automotive industry. This might include the development of renewable energy capacities, better and smarter grids based on decentralized and renewable energy production and improved battery and electric motor engine technologies. The main barrier remains the energy density of the battery which limits the range of all-electric battery planes significantly compared to hydrogen and synthetic fuel. To get the full benefits of an all-electric plane, an entirely new aircraft design that leverages distributed electrified propulsion is required. Similar to

Table 2 Economic evaluation of aviation power and fuel technologies

	Current cost per kWh	Expected cost per kWh in the year 2035	Investment for Aircraft Technology	Investment for Infrastructure
Kerosene	0.03 to 0.07 USD	0.07 to 0.10 USD	Low	Low
Biofuel	0.07 to 0.25 USD	0.05 to 0.08 USD	Low	Med
Synthetic fuel	High, unclear	0.05 to 0.06 USD	Low	High
Hydrogen (produced with renewable energy)	0.10 to 0.14 USD	0.03 to 0.05 USD	High (combustion) Very high (fuel cell)	High (combustion and fuel cell)
Electricity and battery (solar)	0.05 to 0.20 USD	0.02 to 0.04 USD	Very high	High
Hybrid: Synthetic fuel plus electricity and battery (50%)	SF: Unclear Electricity: 0.05 to 0.20 USD	0.03 to 0.5 USD	(very) high	High

Sources: Magone et al. (2021); Thomson et al. (2020); van Bentem (2021)

hydrogen, new commercial aircraft models for all-electric aviation need to be developed first, which requires large research and development costs by plane manufactures into electric motor engines and battery technology as well as a high investment by airlines to renew their fleets.

Hybrid aircraft models based on the usage of 50% synthetic aviation fuel and 50% power usage from battery-powered electricity lead to an average energy cost of synthetic aviation fuel and electricity, which will be in the range of 0.03–0.5 USD per kWh and thus be cheaper than kerosene today. The investment into the infrastructure is similar as it is for all-electric planes. But the investment cost for new aircraft technologies might be slightly smaller compared to hydrogen fuel cell and all-electric aircraft. This is caused by two factors: first, the required battery is much smaller than it would be for an all-electric plane of the same size, which significantly lowers the take-off weight of the aircraft and therefore also lowers the requirements for a new aircraft design. Second, the combination with a combustion engine allows optimal and efficient power supply depending on the flight situation (take-off, landing). Since combustion engines are already the dominant propulsion systems in aviation, the aircraft design does not need to change fundamentally to become a hybrid plane (Table 2).

2.3 Acceptance and Safety Evaluation

With regard to customer acceptance and security of the introduced technologies, the evaluation becomes more difficult, since most of the technologies are not yet available on the market and therefore lack customer experience. Looking at the different aviation fuels, including *kerosene, biofuel*, and *synthetic fuel*, customer acceptance and safety are likely to be the same for all of them (as long as their effectiveness has been proven), since the aircraft technology itself remains the same and therefore does not influence the customer experience of flying compared to today. With regard to kerosene as a fossil fuel, customer acceptance could even decrease in the near future as it has the highest impact on the greenhouse gas effect of all mentioned fuels, while sustainability is becoming an increasingly important topic for travelers. Similar assumptions could be made for biofuel, since it is to some extent competing with agriculture, which in most peoples' minds could compromise the idea of sustainability.

Regarding new aircraft technologies, such as hydrogen (combustion and fuel cell) and all-electric battery planes, one can only make assumptions about customer acceptance and safety. From an ecological perspective, the assumption is that lower emissions and less environmental impact always increases customer acceptance, since sustainability is becoming more and more important to almost everyone who is traveling the world. Considering safety, it is hard to guess how customers would react to new technologies. Since new technologies always involve change at the beginning, and change is usually associated with uncertainty, it is therefore important to ensure airworthiness and safety in commercial flight operations at all times. Of course, new aircraft models are not certified until they clearly meet today's safety standards and all test flights and safety checks have demonstrated that functionality and safety are always warranted. In addition, it is also clear that a change in aircraft technology should not decrease travel comfort compared to today, as this would negatively influence customer acceptance.

2.4 Time to Market and Future Development

Similar to acceptance and safety, it is hard to tell what the future will bring and how fast it will be delivered. With regards to bio- and synthetic fuel, the picture of the future can already be painted better than the one for hydrogen, fuel cell, and all-electric aircraft. While biofuel is already on the market and has proven its reliability as aircraft fuel (since 2018, airlines including Lufthansa, KLM, and others already use some degree of biofuel for commercial flights), synthetic fuel is very likely to do so as well in the next 5 years. The plan is to be on the market from 2023 onward and to be competitive with today's fuel prices by 2035. Also, biofuel needs more investment to achieve economies of scale in order to be competitive with today's fuel. While biofuel is to some extent competing with agriculture, synthetic fuel is not competing with agriculture. This could lead to investors and decision makers in the aviation industry focusing their attention on synthetic fuel rather than biofuel in the near future.

Concerning disruptive technologies, the time to market is highly speculative and can be compared to looking into a crystal ball or reading coffee grounds. While easyJet and Wright Electric announced their 186-seat all-electric battery aircraft for

2030, experts have doubts as to whether battery technology and the required infrastructure will be sufficiently developed by then. Similar to Airbus, who announced their hydrogen combustion aircraft models for 2035. To meet this schedule, the new aircraft types would have to be ready for the market (tested, safe, and certified) by then and the necessary infrastructure for the production, transport, and storage of liquid hydrogen would have to be in place as well. This means, that as promising as the new technologies may be, the timelines may still be overambitious. However, since the new technologies will also be economically profitable in the future and the problem of climate change will likely become even greater, the new technologies will sooner or later become established in the market. The exact point in time, and whether hydrogen or batteries will become the dominant driving force, is unfortunately very difficult to estimate at the present time. According to our evaluation, it is very likely that hydrogen for the combustion engine (Airbus approach) will prevail first, since aviation could continue to rely on the proven combustion engine for aircraft, which means less investment and less time to market. In addition, one could profit from the development of other industries in the field of liquid hydrogen, which could accelerate the time to market. Compared to batteries, hydrogen also offers faster refueling processes and a much higher gravimetric energy density, which increases aircraft ranges and decreases the aircraft weight significantly.

2.5 Summary of Evaluation

After the assessment and evaluation of all relevant current and future technologies for sustainable aviation, it is time to put everything together to get an overview as a comparison. The following graph shows all current and new aviation technologies for medium and long-distance flights with their potential to increase sustainability (based on emissions saved) on the x-axis, and with the required amount for investment into new infrastructure and new aircraft models on the y-axis. Generally, the determination of the investment cost for aircraft models and infrastructure is also highly dependent on future demand growth assumptions. Following the assumption of the industry that air transport could be three times higher compared to 2019, the investment cost for the total transition would be much higher compared to a situation with lower growth levels. Hence, providing exact numbers about investment costs is nearly impossible. As a result, we make relative comparisons about the investment cost for new infrastructure and aircraft models, as illustrated in Fig. 5.

Looking at Fig. 5 clearly illustrates the relationship between required investments and potential emission saving. The higher the potential emission saving, the higher the required investment cost for infrastructure and new aircraft models. It also shows the different pathways one can choose: emission reduction, CO_2 neutral (net zero), (absolute) zero CO_2 emissions, and absolute zero GHG emissions at all. Biofuel and synthetic fuel offer the greatest potential in the short and medium run, since both do not require new aircraft models that need to be developed first. Nevertheless, both types of fuel require significant investment into infrastructure to increase the

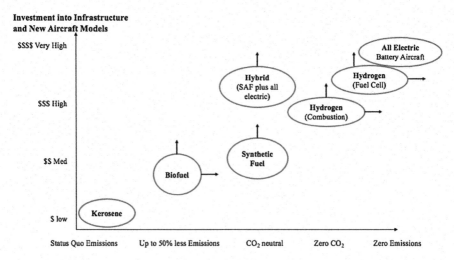

Fig. 5 New aviation technologies—Investment needed and emissions saved. Own illustration, inspired by Thomson (2020)

efficiency of production to achieve competitive prices. This could be realistic between 2030 and 2035. For hydrogen-powered aircraft, new infrastructure for hydrogen production, storage, and distribution is required but also new aircraft models that are capable of transporting sufficient amounts of liquid hydrogen. Both require time, a lot of investment and also highly dependent on the development of other industries that could make use of hydrogen in the future. These circumstances make it difficult to assess when and in what form hydrogen-powered aircraft will play an important role in sustainable flying. The same accounts for fuel cell-based, hybrid (SAF and all-electric) and all-electric aircraft. However, hydrogen for the internal combustion engine could still be superior to fuel cells, hybrids, and all-electric aircraft in terms of investment required and time to market, since the internal combustion engine and its usability for aircraft is already developed to a very high level and thus does not require new aircraft propulsion technologies, such as all-electric propulsion.

Based on this evaluation we believe that biofuel and synthetic fuel will be important drivers to increase sustainability in aviation in the next 10–15 years, while disruptive technologies such as hydrogen for combustion, hydrogen fuel cell and all-electric aircraft are very likely to play an important role earliest from 2035 onward, with some advantages for hydrogen in combustions to be available in the market prior to the other technologies. Given the current development in carbon pricing and incentivizing clean energy sources, it is also very likely that all technologies will be superior in terms of energy cost compared to today's kerosene. Table 3 summarizes and compares all technologies and their evaluation criteria.

Technology Assessment for Sustainable Aviation

Table 3 Evaluation summary of aviation power and fuel technologies

	Negative impact on climate	Economy	Acceptance and safety	Time to market
Kerosene (100%)	Very high impact: Emissions of CO_2, water vapor, NO_X, and other GHG emissions (e.g., soot particles)	Today: Benchmark Future: Potential efficiency gains but also increased CO_2 pricing	High, proven, and most used aviation fuel for decades	Existent, mature
Biofuel (100%)	Moderate impact: Up to 80% less net GHG emissions than kerosene but still radiative forcing	Today: Not competitive Future: Will be similar to kerosene today	Proven—no acceptance or safety problems anticipated	Existent, early stage High investments required
Synthetic fuel (100%)	Moderate impact: Up to 80% less net GHG emissions than kerosene, but still radiative forcing	Today: Not competitive Future: Will be similar to kerosene today	New—But no acceptance or safety problems anticipated	Not existent, expected in 2023 High investments required
Hydrogen (100%)	Low impact: Absolute zero CO_2, Net Zero GHG emissions but still radiative forcing (lower than for kerosene and SAF)	Today: Not competitive Future: Will be cheaper than kerosene	New Aircraft Model: Acceptance and safety unclear	Not existent, expected in 2035 Very high investments required
Electricity (solar) and Battery (100%)	No impact: Absolute zero Emissions during flight No radiative forcing	Today: Not competitive Future: Will be cheaper than kerosene	New Aircraft Model: Acceptance and safety unclear	Not existent, expected in 2030 Very high investments required
Hybrid: SAF plus Electricity and Battery (50%)	Moderate to low impact: 95% less net GHG emissions than kerosene, if battery covers 50% of power usage but still radiative forcing	Today: Not competitive Future: Will be cheaper than kerosene	New Aircraft Model: Acceptance and safety unclear	Not existent, expected in 2030 Very high investments required

Source Table 3: based on previous Tables 1 and 2

3 Supportive Technologies for Sustainable Aviation

3.1 Fuel and Energy Efficiency

3.1.1 Air Navigation Service Provider Optimization Potentials

Technically supported and feasible operational improvements in air traffic control service providers also have the potential to reduce aviation emissions. Two examples should be highlighted: The Single European Sky and Continuous Descent Approaches.

For more than a decade, the EU is working on the *Single European Sky Initiative*. The goal is to develop a modern air traffic management system for Europe, which is designed to improve the environmental performance of aviation. By modernizing the management of the European airspace, more efficient flight paths with fewer detours would be established. The EU estimates that this reorganization could result in up to a 10% reduction of air transport emissions.

The European Commission proposes various actions to ensure safe and cost-effective air traffic management services, including "strengthening the European network and its management to avoid congestion and suboptimal flight routes, promoting a European market for data services needed for a better air traffic management, streamlining the economic regulation of air traffic services provided on behalf of Member States to stimulate greater sustainability and resilience and boosting better coordination for the definition, development and deployment of innovative solutions" (European Commission, 2020). This comprehensive and complex reform has so far been thwarted by a lack of consensus due to higher weighted national interests such as control over national airspace sovereignty.

Continuous Descent Approaches (CDAs) have the potential to significantly reduce fuel burn and noise impact. Arriving aircraft are kept longer at their cruise altitude compared to conventional approaches. The air navigation service provider (ANSP) makes them descend as late as possible in a continuous descent to the runway at near idle thrust with no level flight segments. CDA procedures are vertically optimized fixed routes. Due to changing traffic conditions and variable noise regulations, the advantages of CDA operation have not yet been fully achieved. CDAs require a high level of transparency between pilots and air traffic controllers, most importantly on the timely provision of information and agreeing on clearances early, thus enabling smooth approaches and avoiding lateral and vertical corrections. Below 3000 ft., CDA can result in up to 20% reduction in CO, up to 25% in HC and up to 35% reduction in NOx as well as in noise reductions of up to 30%. So called dynamic CDA approaches can result in further reductions of around 15% in noise, 11.6% reduction in NOx emission, and 1.5% reduction in fuel burn (Alam et al., 2010; Brooks, 2006).

3.2 Ground Services

3.2.1 Electrification and Automation of Ground Handling

The transition from conventional combustion motors to electric or alternate fuel vehicles is also underway among ground service providers. Ground handling agents, such as Dnata and Swissport have set targets for the number of electric vehicles in their fleets. Electric ground servicing equipment (GSE) produces fewer CO_2 emissions compared to conventional GSEs, although the difference between the two varies depending on which parts of manufacturing and lifecycle are considered. Key performance indicators (KPIs) for ground handling activities are punctuality, minimal use of resources, predictability, minimal damage to aircraft, and minimal personnel incidents and accidents. One specific method for improving these indicators is automation. AI is more effective than humans in coordinating large and complex systems. Precise sensors allow for a more exact and timely execution of processes, thus limiting efficiency losses. Yet, no airport currently uses completely automated ground handling systems since fully automated ground handling systems are still in the invention phase. Any system that is implemented at an airport must be tailored specifically to the aircraft types serviced by the ground handler, the number of serviced aircraft, the airport layout, and the target turnaround time. Therefore, also, the configuration of a GSE fleet is specific to the airport. It is dependent on the size of the airport, the desired turnaround time, aircraft type, and the national or airport-imposed regulations of GSEs, which in combination constrains the rapid implementation of electric GSE.

3.2.2 Engines-Off Taxiing

About 60% of all airport emissions are generated in the aircraft landing and take-off (LTO) cycle and thus present a large reduction potential for airports. Engines are not designed for taxiing around airports; they waste a substantial amount of fuel in doing so and produce unnecessary emissions, air pollution, and noise. An engines-off taxiing system would therefore be advantageous for various stakeholders. Benefits include fuel savings for airlines, a more predictable system with fewer delays for airports, and an overall reduced financial risk from incidents and accidents. Four approaches to implementing engines-off taxiing are currently being discussed: operation towing, electrically powered landing gear, self-driving towing vehicles, and aircraft towing systems (ATS) (Morris et al., 2015).

Operation towing is the towing of aircraft by human-driven vehicles. Few logistical and operational changes are required for implementation of operation towing, as human drivers are available, and vehicles can be purchased. Having an additional person responsible for the towing would improve safety and reduce the workload of pilots, who can then concentrate on other tasks instead of on taxiing. This system would require significantly more coordination in the form of vocal communications, which is problematic given the potential for miscommunication and frequency congestion (blocked radio frequencies) common at larger airports.

Electrically powered landing gear shifts the responsibility for taxiing back to the pilot to reduce surface traffic. The solution is technologically limited, as the auxiliary power unit (APU) alone cannot provide enough power to move the aircraft, and aircraft would have to be retrofitted with additional electric engines.

Fully automated or remote-controlled electric tugs offer an alternative solution. Options include taxi-bots, piloted by a tug driver from the flight crew for the pushback segment of departure or self-driving tugs which operate on predetermined routes.

Rail-based aircraft towing systems (ATS) are composed of three sub-systems: a taxiway channel, a pull car, and the managing software. The taxiway channel is built into the taxiways and houses the pull cars. The pull car is positioned on the taxiway as soon as the aircraft vacates the runway. Aircraft taxi onto the pull car, which automatically connects itself to the nose gear. Taxi operations are subsequently controlled by automated software.

Both electric ground vehicles and engines-off taxiing have the potential to reduce emissions in the airport environment. However, there are still open questions regarding technical implementation and regulation, which delay the implementation (see also Alonso Tabares & Mora-Camino, 2017).

4 Conclusion

4.1 Conclusions and Outlook on Technology Development

In the short term (until 2030), there will probably be no technological breakthrough in the form of a game changer. By then, however, alternative fuels, including biofuels and synthetic fuels, will gain in importance, as they are literally the only way to improve sustainability and reduce emissions in air travel (apart from flying less) in the short run. While biofuel already exists on the market and is also already used by numerous airlines (such as KLM, Lufthansa, SAS, and British Airways) for commercial flight operations (mixed with 50% kerosene), synthetic fuel is not yet available on the market. Both forms of fuel have the advantage of being compatible with today's aircraft and combustion engines, which means that investments and time-intensive development cycles for new aircraft models are not required. Additionally, using alternative fuels will result in aircraft having the same ranges as they have today with kerosene. As a result, airlines have rather low conversion costs (flight schedules and aircraft models can remain the same, as ranges and functionality remain the same as for kerosene).

Comparing synthetic fuel with biofuel, one can say from an economic perspective that biofuel will still have a price advantage at the moment and during the next 3–5 years. How the development continues after that depends strongly on investments in the respective infrastructures for the production of the fuel types. From an environmental perspective, synthetic fuel brings other advantages. The CO_2 emissions produced during combustion are all part of a natural CO_2 cycle and are generally said to be 50% less than conventional kerosene. Compared to a blend of 50% biofuel and kerosene, this shows significantly improved emission levels. Further advantages of synthetic fuel compared to biofuel are its scalability for global aviation fuel demand (which is critical with biofuel due to the required amount of land use) and the fact that it does not compete with agricultural land.

Based on these considerations, we believe that synthetic fuel (if investments are taken and scalability is increased) will replace biofuel and a large share of kerosene for short and long-distance flights in the medium run (until 2035) based on environmental and economic reasons. An increased use of synthetic fuel in aviation has a great potential to lower emissions significantly for both, short and long-distance flights. Therefore, we see a big potential for synthetic fuel in the next 10–20 years. Depending on the development of other technologies, carbon pricing measures and the scalability investments for synthetic fuel, it is also very likely that a high demand for synthetic fuel will continue even after the next 15–20 years. Especially if hydrogen and all-electric or hybrid aircraft are not ready for commercial use, synthetic fuel will remain the only sustainable alternative. However, there are still emissions (including other GHGs than CO_2) produced with synthetic fuel too, which should be considered when sustainability measures are taken in the aviation industry, as these emissions still increase radiative forcing which also increases global warming effects.

In the long run (2035 onward), new and disruptive aviation technologies could slowly gain importance. If easyJet and Airbus meet their timelines, the first battery-powered aircraft for short-haul routes (up to 600 kilometers) should be in service from 2030, and the first hydrogen powered aircraft based on hydrogen combustion for short- and medium-haul routes (up to 3600 kilometers) from 2035. From our point of view, both timelines tend to be very ambitious. Not only the required aircraft models need to be developed first and have demonstrated their safety and functionality for commercial aviation, but also new infrastructures for the required energy sources (hydrogen, electricity, and battery) need to be developed first. A not to be underestimated advantage for the development of hydrogen infrastructure and new power grids as well as batteries are the positive spillovers that can come from other industries (Automotive Industry, Power Train Industry, Energy and Power Industry, etc.). This makes it even more difficult to predict the development and its speed in the field of hydrogen and battery technologies. However, both technologies will play an important role in the field of mobility in general. Similar to the car and truck industry, it is expected that batteries will be used for short distances (based on superior energy efficiency) in aviation, while hydrogen could be used for longer distances (due to the higher energy density per weight). Both, hydrogen for combustion and battery-powered aircraft would be superior in terms of sustainability compared to synthetic fuel, since both technologies have no CO_2 emissions at all. Battery-powered planes do not even have any kind of emissions and hydrogen for combustion only emits water vapor and NO_X. Based on these considerations we believe that all-electric battery planes will be a viable option for short-haul flights sometime after 2030. For medium and long-distance flights, hydrogen for combustion will be superior to batteries based on the energy density and weight of hydrogen compared to batteries. Since Airbus announced new aircraft models only for medium distance flights (up to 3600 km), synthetic fuel will potentially still be the only option to achieve higher sustainability for long-distance flights as well. If the hydrogen infrastructure is in place and Airbus keeps track with their timeline, we expect commercial hydrogen aircraft sometime after 2035 in the market. Finally, it should also be noted that a transition to battery and hydrogen aircraft on a global scale in aviation will take at least another decade, in which synthetic fuel might still be the best option for aviation to be more sustainable.

With regard to a very far future, the development becomes even more uncertain. In particular, battery technologies and advanced fuel cells, which can generate enough power to operate commercial passenger aircraft also for medium and longer distances, do not exist at the moment (this also applies to short-haul commercial flights at the moment). For battery-powered aircraft, but also for fuel cell-based aircraft, this means that it is very difficult to estimate when these technologies will actually play a decisive role in air transport sustainability. From an ecological perspective, both technologies have the greatest potential to reduce emissions to (almost) zero in air traffic. Both technologies emit no CO_2 at all, while the fuel cell produces only water vapor as exhaust gas, a battery produces no emissions at all. Both technologies are most likely not ready for long-distance flights before 2050.

4.2 Implications and Recommendations to Achieve Sustainability in Aviation

Since the current situation regarding the future is still associated with a great deal of uncertainty, it is difficult at the moment to formulate really clear recommendations for action. Nevertheless, we identified some fields for action. Some of these recommendations for action already relate to today, while other actions will only gain relevance as time progresses.

Speaking of today, aviation executives would be well advised to start collaborations and partnerships with alternative fuel producers, including bio- and synthetic fuels, to secure direct access to these fuels and thus price advantages for the future. Direct access to alternative fuels could lead to strategic competitive advantages in the medium term, based on unique sustainability and cost benefits (assuming effective carbon pricing for aviation and incentives for CO_2 neutral fuels). Additionally, aviation executives from airlines, but also from infrastructure relevant companies (airports, ground handling, suppliers, etc.), should foster collaboration paths with policy makers to establish a fruitful and stable environment for the development of SAF, which might include financial grants and incentive structures but also public campaigning to increase customer awareness for the use of SAF.

To increase the knowledge about future opportunities in aviation based on technological transitions, we advise aviation executives to explore the economic and ecological potential of new technologies (hydrogen, all-electric battery): either through own research or through collaboration with technology developers (as easyJet does, for instance, with Wright Electric). Both technologies, hydrogen and all-electric (also hybrid with all-electric and synthetic fuel) aircraft have the potential to build strategic competitive advantages with regard to sustainability and cost. Assuming an effective and meaningful carbon pricing mechanism for aviation after 2030 plus incentive structures for clean energy sources, hydrogen and electricity will have clear cost advantages after 2035, most likely also compared to synthetic fuel, since electricity and hydrogen are absolutely carbon free, which is not the case for synthetic fuel. Executives are well advised to carefully monitor the development of future aviation technologies and to allocate resources and investments to pave the way for a successful transformation today already.

Short-term Recommendations:

- Start collaborations or partnerships with SAF producers as early as possible.
- Support the establishment of a stable and fruitful policy environment for SAF production.
- Continue to realize fuel efficiency gains wherever possible.

Long-term Recommendations:

- Carefully monitor the developments of hydrogen, battery all-electric, and battery hybrid aircraft with respect to sustainability and ecological effects.
- Prepare for heavy future investments in new infrastructure and new aircraft models.

References

Airbus. (2021). *Company Homepage and Newsroom*. Retrieved April 22, 2021, from https://www.airbus.com/newsroom/press-releases/en/2020/09/airbus-reveals-new-zeroemission-concept-aircraft.html

Alam, S., Nguyen, M. H., Abbass, H. A., Lokan, C., Ellejmi, M., & Kirby, S. (2010, October). *A dynamic continuous descent approach methodology for low noise and emission*. In 29th Digital Avionics Systems Conference (pp. 1-E). IEEE.

Alonso Tabares, D. & Mora-Camino, F. (Eds.). (2017). *Aircraft ground handling: Analysis for automation*, Denver, United States, American Institute of Aeronautics and Astronautics.

Brooks, J. (2006, September 20–21). *Continuous descent arrivals*. In ICAO workshop on aviation operational measures for fuel and emissions reductions.

Business Traveller. (2020). *EasyJet and Wright Electric Case*. Retrieved April 22, 2021, from https://www.businesstraveller.com/business-travel/2020/01/31/easyjet-partner-plans-to-test-electric-aircraft-in-2023/

Dial, S. (2011). *A plot of selected energy densities (excluding oxidizers)*. Retrieved February 23, 2021, from https://upload.wikimedia.org/wikipedia/commons/c/c6/Energy_density.svg

European Commission. (2020). *Single European Sky: For a more sustainable and resilient air traffic management*. Retrieved March 11, 2021, from https://ec.europa.eu/transport/modes/air/news/2020-09-22-ses-more-sustainable-and-resilient-air-traffic-management_en

Heyne, J., Rauch, B., Le Clercq, P., & Colket, M. (2021). Sustainable aviation fuel prescreening tools and procedures. *Fuel, 290*, 120004.

Lee, D. S., Fahey, D. W., Skowron, A., Allen, M. R., Burkhardt, U., Chen, Q., et al. (2021). The contribution of global aviation to anthropogenic climate forcing for 2000 to 2018. *Atmospheric Environment, 244*, 117834.

Liu, H., Huang, Y., Yuan, H., Yin, X., & Wu, C. (2018). Life cycle assessment of biofuels in China: Status and challenges. *Renewable and Sustainable Energy Reviews, 97*, 301–322.

Magone, L. G., Barker, A., & Peltz, L. (2021). *Life cycle assessment of producing synthetic fuel via the Fischer-Tropsch power to liquid process*. In AIAA Scitech 2021 Forum (p. 0261).

Morris, R., Chang, M. L., Archer, R., Cross, E. V., Thompson, S., Franke, J. L., Garrett, R. C., Malik, W., McGuire, K. and Hemann, G. (2015) *Self-driving aircraft towing vehicles: A preliminary report, AAAI*.

NASA. (2021). *Boundary layer ingestion propulsion*. Retrieved February 23, 2021, from https://www1.grc.nasa.gov/aeronautics/bli/

Neste. (2021). *Company homepage*. Retrieved July 14, 2021, from https://www.neste.com/about-neste

Pavlenko, N., & Searle, S. (2021). *Assessing the sustainability implications of alternative aviation fuels*.

Quaschning, V. (2021). *Specific carbon dioxide emissions of various fuels*. Retrieved February 18, 2021, from https://www.volker-quaschning.de/datserv/CO2-spez/index_e.php

Thomson, R., Weichenhain, U., Sachdeva, N. & Kaufmann, M. (2020). *Hydrogen: A future aviation fuel?* Roland Berger GmBH Munich. Retrieved February 01, 2021, from https://www.rolandberger.com/nl/Insights/Publications/Hydrogen-A-future-fuel-for-aviation.html

van Bentem, K. (2021). *Techno-economic analysis of sustainable aviation fuels by using traffic forecasts and fuel Price projections: A case study at TUI aviation*.

Zhenli, C. H. E. N., Zhang, M., Yingchun, C. H. E. N., Weimin, S. A. N. G., Zhaoguang, T. A. N., Dong, L. I., & Zhang, B. (2019). Assessment on critical technologies for conceptual design of blended-wing-body civil aircraft. *Chinese Journal of Aeronautics, 32*(8), 1797–1827.

Perceptions of Flight Shame and Consumer Segments in Switzerland

Philipp Gunziger, Andreas Wittmer, and René Puls

Abstract

- The flight shame movement gave impetus to a critical reflection of air travel in the context of global sustainability debates, fostering the idea of using trains instead of planes for short-haul distances.
- Forty-three percent of a representative study perceive flight shame as a positive development, while 33% are clearly negative on the subject, which implies a controversial topic. 21% do, however, not know what flight shame is and what it comprises.
- Based on attitudinal and behavioral segmentation criteria, four consumer segments in air travel were identified. Nearly 40% of consumers can be considered eco-friendly, while around a fifth shows inconsistencies between eco-friendly attitudes, but contradictory behavior. A third of consumers can be seen as non-eco-friendly. These four consumer segments predominantly differ in their political orientation and the number of cars per household.
- The behavioral characteristics of these segments show an overall modest desirability for train use across all consumer groups (58%), implying a general willingness of customers to use trains for short-haul distances (≤ 700 km).
- In the light of sustainability and flight shame, consumers generally demand conscious and reasonable air travel behavior, which entails a renunciation of ultra-low ticket prices and a reduction of short-haul flights, by simultaneously enhancing railway infrastructure and connections to provide a viable and more environmentally friendly alternative.

P. Gunziger (✉)
University of St. Gallen, St. Gallen, Switzerland

A. Wittmer · R. Puls
Center for Aviation Competence, University of St. Gallen, St. Gallen, Switzerland
e-mail: andreas.wittmer@unisg.ch; rene.puls@unisg.ch

© The Author(s), under exclusive license to Springer Nature Switzerland AG 2022
J. L. Walls, A. Wittmer (eds.), *Sustainable Aviation*, Management for Professionals,
https://doi.org/10.1007/978-3-030-90895-9_3

- Long-haul flights are, in contrast, predominantly seen as an indispensability, where the focus should lie on technological improvement in order to make air travel environmentally compatible.
- Regarding decision criteria for using modes of transportation, price, travel time, and CO_2 emissions were identified as critical factors. Interestingly, as long as journeys are within a critical time threshold, trains are preferred although the corresponding price is higher than for a trip by plane.

1 Status Quo of Air Travel in Switzerland

Consumer preferences and the resulting behavior are major drivers for sustainability shifts in any industry (Bask et al., 2013). Therefore, a focus on consumer preferences and behavior can provide important insights into the future development of the aviation industry. To understand consumer preferences, the focus in this chapter lies on exploring attitudes toward flight shame and resulting behavior regarding air travel and sustainability. As the study design and database stem from Swiss consumers, we explicitly concentrate on Switzerland in this chapter. Before taking a deeper perspective into the topic, we take a brief look at the facts concerning air travel in Switzerland.

1.1 Air Travel Statistics

Air travel has seen, apart from a few shocks, continuous growth in passenger numbers for the last 50 years. The number of flights taken and overall air traffic volume have constantly increased and are estimated to continue to do so in the next years (Graham & Metz, 2017). In Switzerland, as highly interconnected and affluent nation, the growth tendency is even stronger. The number of air trips per person and year has risen by 43% within 5 years, whereas in 2019 an average citizen covered 9'000 kilometers by plane annually and travels twice as much per plane compared to consumers from Germany or Austria (Intraplan, 2019). On a per-capita basis, Swiss air travelers are therefore among the biggest emitters globally (Stalder, 2017). Further of relevance is the fact that air travel is not evenly pronounced in societies. Statistics of air travel reveal that rather a small number of passengers is responsible for most flights. With small differences, studies concerning flight consumption show that approximately 80% of all trips are undertaken by only 20% of consumers (Kroesen, 2013). Those numbers and growth rates are indicators of the hypermobile society in which we live in, with pervasive consumerism extending to our mobility behavior. However, flying patterns differ widely among different consumer segments, and so do related attitudes and values.

1.2 Conflicting Attitudes

Taking up attitudes and perceptions of consumers, climate change and sustainability concerns are key topics. These debates are in harsh contrast to the growth outlooks of aviation, as climate change and global warming are major sources of worry at the same time. In Switzerland, climate change and related problems are among the most important concerns. According to a survey on the concerns of the nation in 2019, environmental protection and climate change surpassed worries about unemployment, personal safety, and new poverty (Credit Suisse, 2019). In a consumer barometer in the field of renewable energy in 2020, strong feelings were revealed on climate change: 53% feel sad about climate change, while 37% are afraid of it, also because 69% agreed that the region in which they live is or will be affected by climate change in the future (IWÖ-HSG, 2020). In the wake of this risen environmental consciousness, air travel has a hard stance. Aviation contributes toward 2% to 2.5% of total global carbon emissions (ATAG, 2020). In Switzerland, emissions caused by aviation contribute to even 12%–18% of total CO_2 emissions, thereby representing a significant share of the country's carbon footprint (The Swiss Parliament, 2019). Such a massive environmental impact raises concerns in the local society and thereby also exerts influence on their behavior. In a WWF study in 2019, 40% of respondents were willing to reduce their number of flights due to environmental reasons, while 27% indicated that they already did so (WWF, 2019). Apart from renouncing, another tendency related to air travel behavior, is using alternatives such as trains. Also in this area, statistics indicate a shift in behavioral patterns, as railways recorded an increase in train trips in the last years. Similar to the ones in Sweden and Austria, Swiss Federal Railways (SBB) quantify an increase of 10% in international connections in 2019, and even a rise of 25% in night trains. Although SBB states that this increase cannot clearly be attributed to the climate change and sustainability discussion, customer surveys indicate that sustainability considerations are gaining in importance (Meier, 2019).

1.3 Behavioral Consequences

Such a change in behavioral patterns is, seen from an ecological point of view, a promising way for future travel. Substituting air travel by train trips can significantly reduce the CO_2 emissions caused by transportation, as trains can save around 80–90% of related CO_2 emissions compared to planes, especially on short-haul routes (Dällenbach, 2020). Railways can therefore be regarded as very energy-efficient and low-emitting modes of transport, and thus offer a promising alternative for resource-intensive air travel.

1.4 Chapter Structure

For those reasons, the focus of this chapter lies especially on the use of trains as a viable alternative for short-haul air travelers. To better grasp the ecological awareness of consumers in air travel, flight shame and its perceptions are viewed in detail in the subsequent chapter. Building on this, different consumer segments and their characteristics, as well as their possible future behavior, are presented in the third intercept. It is important to mention that air travel behavior in this chapter solely focuses on leisure travel. Business travel is exempted because underlying attitudes and motives are considerably different. In the ensuing section, decision criteria for the two modes of transportation—plane or train—are elaborated to better understand the decision-making process of travelers. At the end of this chapter, findings are condensed to present consumer expectations for future air travel in the context of sustainability.

The database for Sections two and three is a representative study of Switzerland about the future of aviation, conducted by the Center of Aviation Competence (CFAC-HSG) in cooperation with the Aviation Research Center Switzerland (ARCS) (Wittmer & Puls, 2019). The aim of the study was to identify the role of aviation with respect to prosperity, wealth, and well-being of Swiss society in the year 2040. To handle the complexity of a variational, future-oriented topic, a scenario- and projection-based online Delphi study was carried out, whereby 3000 panelists had to assess the desirability of different projections on a 10-point Likert scale. In this Delphi study design, each panelist provided an initial assessment concerning the preference of a projection. This initial assessment was followed by presenting the average score of the peer group, allowing the participants to reconsider their first assessment and provide a second score, or keeping the initial one. As the consensus process and the score variations between the two assessments were not part of this study, only scores from the first answer were analyzed, in order to obtain unbiased responses.

The fourth section of this chapter is based on a survey study with Swiss students. Due to those study designs, presented findings are based on Swiss consumers, which is why this chapter is focused on Switzerland. Although Swiss consumers generally have an above-average air travel consumption, the insights gained in the study can to a certain extent be revealing for and transferred to other highly industrialized countries in Europe as well.

2 Ecological Awareness of Consumers and Their Perception of Flight Shame

2.1 What Is Flight Shame?

Increasing environmental concerns in society are becoming more and more incompatible with resource-intensive modes of transportation such as air travel. Already 10 years ago, the impasse of deeply embedded air travel practices on the one hand

and collective climate change consequences of such practices, on the other hand, were labelled as the flyers' dilemma. In that context, results of conducted studies showed that consumers were aware of the negative consequences of air travel, but the step from attitudes to behavioral changes remained intractable. The nature of this dilemma reappeared recently and was taken on a higher and more visible level. The discourse about environmentally harmful behavioral practices culminated in the so-called flight shame movement, which came up in 2017 and gained popularity especially because of the Fridays for Future movement and their prominent leader Greta Thunberg. Initiated by five Swedish celebrities, the original demand was to give up air travel for the sake of environment (Patel, 2020). Flight shame, adapted from the original Swedish word *flygskam*, refers to the guilt people feel from using airplanes as a mode of transportation (Abend, 2019), and to the feeling of being embarrassed or ashamed because of the environmental impact of using planes (Henley, 2019). Flight shame, which was quickly spread throughout the world through extensive media coverage, induced a field of force in which the necessity and justifiability of air travel are debated (Gössling et al., 2020). A central demand resulting from flight shame debates is to reduce flying in order to use more sustainable and less resource-intensive means of transport and thereby reducing one's carbon footprint. Especially the use of trains instead of planes is a popular change in mobility behavior, for which even a separate hashtag—*tagskryt* (train bragging)—has been created and is accompanied by waves of social media posts showing people travelling by train.

The global media coverage of flight shame created an intensive discourse, with many diverging reactions. On the one side, sympathizers see flight shame and its consequences as a necessary signal in times of climate change, where everyone must rethink his or her own behavior, especially in air travel. In times where governments declare climate emergencies, there is, according to a growing number of people, no justification for flying anymore. On the other side, critical opinions argue that flying is an essential part of today's globalized and interconnected world. Hence, society should see flying as indispensable and shift the focus on environmental compatibility to new technologies rather than calling for the abandonment of flying.

The actual impact of flight shame can be gauged by looking at travel data statistics. These numbers indicate that calls for flying less seem to be effective, at least on a local level. According to a survey released in May 2019 by Swedish Railways, 37% of respondents chose to travel by train instead of plane where possible, compared to 20% at the start of 2018 (Gerretsen, 2019). This resulted in a rise of train journeys of 5% in 2018, and an increase of 8% in 2019, whereas business trips by train increased by 12% even (Henley, 2019). Also, ÖBB, Europe's largest international passenger rail company, has seen a 10% growth on its lines in 2019 (Abend, 2019). This is supported by numbers of the Swiss Federal Statistical Office, which state that in 2019 fewer passengers in Switzerland took the plane for destinations within Europe (e.g., Rome -9.7%, Hamburg -6.0%, Venice -2.5%, Paris -2.3%). However, this reduction could not be observed on long-haul flights. For typical holiday destinations like Punta Cana, Sao Paulo, Marrakech but also New York and Montreal, an increase in flights could be recorded. This suggests that consumers change their flying patterns, but do not reduce their air travel generally (Ehrbar, 2020). One of the very few studies already conducted regarding flight shame supports this tendency, highlighting the importance of flight shame with respect to social norms. As these have changed in the light of flight shame, extensive air travel has to be individually justified, questioning long-established views of air travel implying social status. Although flight shame has already penetrated society, consumers still do not report a significant change in travel behavior (Gössling et al., 2020). The impact of flight shame is further relativized when looking at the issue on a global scale. Environmentally conscious behavior in aviation remains to be a mostly European-driven phenomenon. In Asia, sustainability concerns regarding air travel are scarce. The rise of middle classes in Asian countries entails a massively growing appetite for international travel, which leads to substantial and continuous growth in air travel. In the United States, air travel continues to be a vital part of domestic travelling, as few alternatives exist in this widespread country and railway infrastructure is still less developed compared to Europe. The same applies to Australia, which in addition does not have other viable alternatives than air travel for international trips.

2.2 Perception of Flight Shame

Flight shame and its underlying ideology are highly controversial. In academic literature, flight shame debates are in a broader context seen as supportive of changes in social norms, which might lead to changes in behavior due to increased social pressure. However, scientific evidence of flight shame and its perceptions are scarce and, apart from a few surveys, mostly inexistent. This raises the question of how people perceive flight shame and, in this light, their own air travel behavior. To explore this issue, the focus is turned back to Switzerland and the representative study on future aviation, introduced in the first chapter. In an open-ended question, panelists of the online Delphi study could freely state their associations with flight shame, which enables an examination of the perception of flight shame.

Before looking at detailed reasonings, the perception of flight shame is evaluated on an aggregate level. As seen in the illustration below, a majority has positive attitudes toward the topic, while a third is critical on the subject. Almost a fifth stated that they do not know what flight shame is and a tiny portion has no opinion on it. This structure reveals a high consciousness for environmental topics, as the largest share of respondents sees predominantly positive aspects in the movement. The impression of a controversial topic can also be confirmed, as a clear counterpart to the advocate section exists. Perhaps surprising is the fact that 21% do not know what flight shame is, contrasting the perception of media coverage, which depicted flight shame as an omnipresent and well-known topic (Fig. 1).

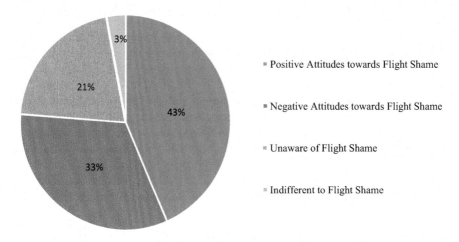

Fig. 1 Perception of flight shame. Own illustration

2.3 Reasonings of Flight Shame Perceptions

In order to gain a deeper understanding of customers and their ecological awareness, the opinions underlying these presented perceptions must explored in detail. The advocates of the flight shame movement put forward as one of their dominant arguments that a more conscious consumption in air travel is necessary. This targets especially binge flying and related short-haul flights for shopping trips or comparable activities. Such practices are increasingly less accepted by the environmentally conscious part of society as flying is not supposed to be a discretionary consumer good. Instead, it should be an appreciated means of transport, allowing people to undertake long journeys and thereby discover new countries, cultures, and broadening one's horizon. Another main reason why these 43% see flight shame as a positive movement is because it fosters the use of alternatives to planes. Especially trains are seen as viable alternative for short-haul distances within Europe, as they are seen as a clearly more environmentally friendly way of travelling. Furthermore, they enable a more conscious travel experience, by being able to explore new landscapes and cities while travelling. The third main argument of flight shame supporters is the criticism of low airfares. The development of decreasing ticket prices is seen critically by this consumer group, as low costs for air travel induces a new, and perhaps artificial, demand. Such low ticket prices, especially for short-haul trips, are not compatible with climate change and increasing environmental problems according to the responses to the open questions. In this regard, environmentally conscious consumers expect actions of airlines and politicians, in order to inhibit the increasing availability of low-cost flights. Such consumer requests related to their perceptions are presented in detail in the fifth subchapter.

The counterpart of the flight shame friendly subgroup is the 33% of flight shame critics. The most common argument of this group is that flight shame is only considered as a fad. They see the flight shame, as well as the Fridays for Future movement as a temporary appearance and as a hype, which will disappear without leaving a substantial impact on society. Thereby, flight shame activists are seen as pure idealists, decoupled from reality. Instead, flight shame critics call for conscious air travel consumption, taking a stance on self-responsibility. Rather than abstaining from flying, efforts have to be undertaken to provide technological improvements that make air travel compatible with the current emission problems. Interestingly, conscious consumption also serves as an argument for this group, but here the call for consciousness is used to delegitimize the flight shame movement and to present it as unnecessary, as individuals have to determine their own behavior. A third main argument was to relativize the negative impacts of aviation regarding climate pollution. Instead of scapegoating aviation, other transport sectors like shipping, cruises or general car traffic need more attention according to respondents of this group.

2.4 Perceptions and Behavior

To bridge the gap between perceptions and behavior, the Delphi study's respondents stated number of taken flights within the last 12 months were compared based on their affiliation to one of those groups of perception. The analysis shows that people who have positive attitudes toward flight shame fly significantly less than individuals who have a negative perception of flight shame. This supports the impression of flight shame sympathizers being more environmentally conscious and friendly, as this group exhibits a self-reported average of 2.14 flights taken within the last year, whereas the flight shame critics take the plane on average 3.05 times per year. In relative terms, flight shame supporters take 43% fewer air trips than critics of the movement. This finding contradicts the well-known phenomenon of *attitude–behavior gaps* (Kollmuss & Agyeman, 2002), where people's actions contradict their (often environmentally friendly) values and attitudes. These two groups, however, seem to act according to their perceptions and attitudes.

However, the fraction of people who have negative attitudes toward flight shame reveal typical denial patterns, which are often used in the context of an existing attitude–behavior gap. These denial patterns serve as excuses to justify their flying behavior. If they truly had the attitude that flying is not a problem, they would not feel the need for justification. This implies that they also show some extent of (hidden) flight shame or at least a general awareness of the negative environmental impact of their flying behavior. Especially the argument of need for climate action in other areas shows analogies with downward comparison, as respondents argue that flying does not have a very negative impact on climate compared to other polluters like cruise ships. This is similar to the denial strategy of exception handling (Juvan & Dolnicar, 2014), where holidays and related flights are seen as part of an exception, where one can allow oneself something extraordinary. Thus, common behavioral rules displayed in everyday life—as sustainable as they may be—do not apply anymore.

3 Consumer Segments and Characteristics

3.1 Segmentation of Consumers in Air Travel

The fragmented perception of flight shame indicates differences in attitudes and environmental consciousness of consumers. To investigate this further, we conducted a cluster analysis. This approach aims to find clear differentiations between customers, in order to identify distinctive consumer groups. Apart from identifying consumer segments, a second goal of cluster analysis was to reveal their respective characteristics and thereby creating an individual profile for each segment. As a third component, cluster analysis can give, based on a behavioral criterium, an outlook for possible behavioral tendencies for these formerly identified consumer segments. As differentiators, two attributes—attitudinal and behavioral—were used to form the segments through cluster analysis. The former reflects perceptions and opinions of concrete phenomena like flight shame, whereas behavioral attributes refer to the stated air travel behavior of consumers. Attitudinal components were chosen because of their widespread use in behavior predicting models. Although limitations like the attitude–behavior gap exist, attitudes are still the most important factor for future-oriented studies and are widely used in research concerning sustainability and environmental awareness due to their temporal stability. As a second differentiating component, behavioral intentions were included in the segmentation process, because they serve as a reference for future behavior. The combination of both enables the examination of congruences or incongruences between attitudes and behavior. In a subsequent step, a set of sociodemographic descriptor variables was used to describe the built segments in further detail. The analysis of the explained segmentation criteria leads to four different customer segments, as mapped below and explained in detail subsequently (Fig. 2).

The largest fraction of consumers, which can be seen by the size of the bubble, is labelled as Non-Eco-Friendly. This is due to the lowest average value on the

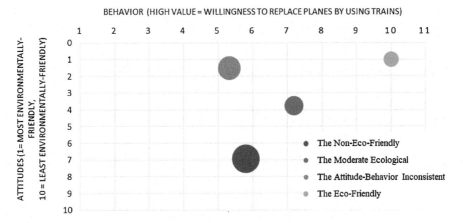

Fig. 2 Cluster score distribution (own illustration)

attitudinal scale, as the CO_2 ticket surcharge on air tickets is mostly not desired by this group. The behavioral variable of this segment shows a less clear picture, as the willingness to use trains instead of planes for short-haul distances (≤ 700 km) is quite evenly spread within the non-eco-friendly groups, including clear opponents as well as advocates of this substitution. This indicates an indifferent and ambiguous behavioral tendency to use trains instead of planes for this cluster. Noteworthy is doubtlessly the size of this cluster, as with 33% it represents the largest consumer segment. However, this can be put into perspective by looking at the other segments in detail. The second largest group, the Attitude-Behavior Inconsistent, exhibit, in contrast to the first one, a high environmental friendliness regarding their attitudes, but are inconsistent when it comes to corresponding behavior, which implies a clear attitude–behavior gap. The willingness to use trains instead of planes is even lower than in the previous Non-Eco-Friendly segment. This group is therefore a clear sign for an existing attitude–behavior gap, exhibiting a divergence between eco-friendly attitudes and corresponding behavior. The third cluster shows a moderate desirability for a CO_2 surcharge, while having a clear desirability for substituting planes by trains. The mean scores support the impression of a distinct profile with pro-environmental attitudes, without being polarized. The last and most distinctive segment is the Eco-Friendly. Scores demonstrate the highest value in environmental friendliness as well as in willingness to substitute planes by trains.

Beyond that, Fig. 5 in Chap. "Technology Assessment for Sustainable Aviation" exemplifies several interesting instances. First, there is only one distinctive profile (the Eco-Friendly) as can be seen in the right upper part of the Fig. 5 in Chap. "Technology Assessment for Sustainable Aviation." All other clusters reveal at least on one of the two variables inconsistencies or an evenly spread distribution of scores, suggesting that there are no opinions or behavioral tendencies carved in stone. Linked to this, there is no counterpart to the Eco-Friendly group, with low environmental attitudes and low willingness for a plane substitution *combined*. Although there is a cluster labelled Non-Eco-Friendly, its characteristics are far less clear-cut than those from the Eco-Friendly. Second, comparing the average score of all clusters, the results reveal a modest pro-environmental attitude (3.72 on the 1–10 scale, with 1 representing the highest environmental friendliness) in total, and also a moderate desirability for using trains instead of planes (6.72 on the 1–10 scale, with 10 representing the highest willingness for substitution). This falsifies the first impression of a less environmentally friendly and conscious consumer base, indicated by the largest non-ecological cluster. Third, there is no cluster with fundamental opposition to using trains instead of planes. This can be seen on the graph because no cluster is below 5 on the behavioral scale. Even the Non-Eco-Friendly cluster expresses on average a moderate desirability for using trains as substitute for planes. The environmentally friendly clusters with their high willingness even signal a complete substitution.

3.2 Profiling Consumer Segments

After revealing four distinctive consumer segments, they will now be profiled in detail to get a comprehensive picture of different consumer types in air travel. Therefore, sociodemographic variables are analyzed and compared between the respective segments. The subsequent table gives detailed information about the sociodemographic variables and their statistical characteristics for each segment, therefore serving as basis for the ensuing profiling (Table 1).

3.2.1 The Non-Eco-Friendly

For the first and largest cluster, age structure indicates a slightly older cluster than the average of the sample, as Gen Z (0–23) are below average, while the share in Baby Boomers (56–73) is higher than usual. In terms of education, the Non-Eco-Friendly have a lower share of academical background while being more firmly anchored in the professional practice. Taking a closer look at the flying behavior, this cluster has taken more flights than any other cluster in the past year. This cluster has the highest share of frequent flyers (>10 flights per year) of all groups, which fits the impression of low environmental concerns. A further indication of an abundant mobility pattern is the fact that the Non-Eco-Friendly own most cars per household, where especially the share of three cars and more is higher than in all other segments. The income structure reveals that this cluster has the highest share of low-income households ($\leq$ 6'000 CHF per month), but also the highest share of high-income households (> 12'000 CHF monthly). Contrary to other research findings, the highly mobile individuals are not only the wealthy ones, but also those from lower income classes. This might allude to two subgroups within this cluster, one as a high-income class, seeing travelling as their personal freedom and unwilling to restrict their behavior on environmental grounds. The other subgroup however is part of the low-income class, seeing flying as their acquired merit, and not willing to refrain from their well-deserved pleasure. The last feature, political party preference, exposes another peculiarity of this segment. Right party affiliation is the highest of all groups (16%) which would suggest a more conservative consumer group. Contradicting in this context is the fact that an even higher proportion (25%) of Non-Eco-Friendly supports left parties, which leaves a split impression of this segment in terms of political orientation.

3.2.2 The Attitude–Behavior Inconsistent

The second largest cluster is younger than the other groups, as the share of Gen Z (0–23 years) is higher than average and the proportion of Baby Boomers (56–73) is the lowest of all segments.

In terms of education, the attitude–behavior inconsistent have a higher proportion of individuals with a degree, while fewer individuals have completed an apprenticeship compared to the previous clusters. Regarding flying behavior, this cluster has the smallest share of non-flyers and the highest percentage of people who flew 6–10 times in the last year, proving the behavioral inconsistency also in this field. Further analyses identify this group as early adopters compared to other segments, meaning

Table 1 Demographics by segment (own representation)

	Non-eco-friendly	Moderate ecological	Attitude–behavior inconsistent	Ecological	Total	Sig.
Cluster size	**33.0%** (*n* = 275)	**22.2%** (*n* = 185)	**27.3%** (*n* = 227)	**17.5%** (*n* = 146)		
Gender						n.s.
Male	46.20%	46.50%	47.10%	46.60%	46.60%	
Female	53.80%	53.50%	52.90%	53.40%	53.40%	
Age (in 2019)						n.s.
Gen Z (0–23)	9.6%	7.6%	14.4%	15.9%	11.7%	
Gen Y (24–42)	45.8%	43.2%	45.0%	44.1%	44.4%	
Gen X (43–55)	23.6%	25.8%	24.7%	22.7%	24.3%	
Baby boomer (56–73)	21.0%	23.1%	15.8%	18.1%	19.5%	
Education						n.s.
Practice	6.10%	4.80%	7.50%	8.20%	6.50%	
Apprenticeship	31.00%	28.60%	20.70%	16.40%	25.10%	
Academia	63.00%	66.40%	71.80%	75.30%	68.30%	
Flights taken in the last 12 months						n.s.
0	19.30%	18.90%	15.00%	24.00%	18.80%	
1–2	44.00%	43.80%	44.00%	45.90%	44.30%	
3–5	24.10%	22.20%	23.80%	19.10%	22.60%	
6–10	7.30%	13.00%	14.50%	10.40%	10.90%	
>10	5.50%	2.10%	2.50%	0.70%	3.10%	
Number of cars in household						***
None	22.20%	13.00%	22.00%	34.90%	22.30%	
1 car	45.10%	54.10%	41.00%	42.50%	45.50%	
2 cars	22.50%	26.50%	30.80%	17.10%	24.70%	
3 cars	8.00%	5.90%	5.30%	4.80%	6.20%	
> 3 cars	2.20%	0.50%	0.90%	0.70%	1.20%	
Early adopter						+
I am always one of the first to buy or use new technologies and equipment.	20.30%	18.80%	21.60%	14.40%	19.30%	
I only start using new technologies and devices when I know what others have experienced with it.	46.20%	47.20%	54.50%	46.80%	48.80%	
I only adopt new technologies and equipment when it is essential for me	33.50%	34.10%	23.90%	38.80%	31.90%	

(continued)

Table 1 (continued)

	Non-eco-friendly	Moderate ecological	Attitude–behavior inconsistent	Ecological	Total	Sig.
personally or professionally.						
Marital status						+
Single	50.60%	43.00%	58.70%	57.00%	52.30%	
Married	40.10%	46.40%	31.40%	30.30%	37.40%	
Divorced	8.60%	10.10%	9.90%	11.30%	9.70%	
Widow/er	0.70%	0.60%	0%	0.70%	0.50%	
Household income						n.s.
Low (0–6'000 CHF)	23.60%	16.80%	22.00%	18.50%	20.80%	
Middle (6'001–12'000)	38.50%	47.60%	47.20%	45.20%	44.10%	
High (>12'000)	17.10%	14.10%	13.20%	16.40%	15.20%	
No specification	20.70%	21.60%	17.60%	19.90%	19.90%	
Political party preference						**
Right	16.20%	9.90%	11.20%	3.10%	11.20%	
Left	24.70%	28.60%	22.50%	42.50%	27.90%	
Middle-right	17.40%	16.20%	18.60%	12.60%	16.60%	
Middle	3.60%	8.70%	8.30%	3.90%	6.10%	
Middle-left	7.70%	11.80%	11.80%	15.80%	11.10%	
No party	18.60%	13.70%	18.50%	12.60%	16.50%	
No specification	6.10%	7.50%	5.40%	3.10%	5.70%	
Other party	5.70%	3.70%	3.90%	6.30%	4.90%	

Note: Significance for Chi-Square distributions: n.s. = no significance, + = moderate significant with $0.05 \leq p < 0.1$, * = significant with $0.01 \leq p < 0.05$, ** = very significant with $0.001 \leq p < 0.01$, *** = highly significant with $p < 0.001$

that members of this cluster are more likely to use or buy new technologies and equipment. Individuals of this cluster are predominantly single, with a comparatively low share of married individuals. Income structure reveals a slightly higher share of low- and medium-income households. Political orientation indicates stronger support for middle parties than other groups. Although this cluster shows a relatively high environmental friendliness on the attitudinal variable, its support for left-wing parties is the lowest of all segments.

3.2.3 The Moderate Eco-Friendly

This cluster is the opposite of the previous cluster in terms of age structure. With the lowest share of Gen Z (0–23) and the highest of Baby Boomers (56–73), it can be regarded as the oldest cluster, with a comparable but more accentuated age structure than the Non-Eco-Friendly. The analogies to that cluster continue in the field of education, where the moderate Eco-Friendly have a slightly lower percentage of academic graduates and an above average share of individuals with an education at an apprenticeship level. It further exhibits the lowest share of households with no cars, while predominantly owning one car per household. Flying behavior shows no peculiarities and corresponds roughly to the average of the sample. The Moderate Eco-Friendly are mostly married, with the lowest partition in single individuals. Further, they are predominantly middle-income households, while low-income households are underrepresented. The political orientation of this cluster shows the highest proportion of middle party support, with a below average right-wing party base.

3.2.4 The Eco-Friendly

The last and smallest cluster is the most recognizable in segmentation variables. Therefore, it is interesting to analyze if these patterns are also apparent in descriptive variables. This seems to be the case: The Eco-Friendly have the highest share of Gen Z and a slightly below-average share of baby boomers, indicating a younger cluster than average. This cluster shows the highest share of individuals with a low level of education, but at the same time the highest proportion of graduates. A clear picture can be drawn in the field of flying behavior. The ecological cluster has the highest proportion of non-flyers, while the majority has undertaken one or two flights in the last 12 months. Frequent flyers (>10) are massively underrepresented in this cluster. This pattern is also evident in the possessed number of cars, where this segment has a share of 35% of households without a car, whereby the quota for the remaining numbers of cars is also below average. The Eco-Friendly further cannot be categorized as early adopters, as they have an above-average proportion of laggards and comparatively fewer early adopters than other segments. Marital status reveals a relatively high share of single individuals and the smallest fraction of married people. Analysis of household incomes reports no peculiarities, as the low-income group is only slightly underrepresented, while the high-income group is only marginally overrepresented. An idiosyncrasy however is evident in political party preference. The ecological group has by far the lowest support for right-wing parties, whereas support for left-wing parties is very strong (43%). Consequently, middle-left party support is also above average, while middle and middle-right party affiliation is lower than in other clusters.

3.2.5 Overview of Swiss Consumer Clusters

Consolidating the profiling with the used descriptive variables, it first must be mentioned that the customer segments show no differences in gender at all. Also, income, generally a powerful differentiator in studies concerning sustainability and ecological awareness, is not significant in profiling segments of this study. Other "classic" demographic variables, such as age and education level, show differences, as well as the number of flights taken, but all of these are not statistically significant. In contrast, the number of cars in households shows highly significant differences between clusters, especially between the Non-Eco-Friendly and the Eco-Friendly. Early adopter and marital status show moderate significant differences, mostly between the moderate Eco-Friendly and the Attitude–Behavior inconsistent. Political party preference as the last descriptive variable has a high discriminatory power,

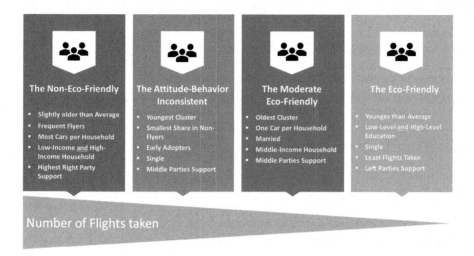

Fig. 3 Cluster characteristics (Own illustration)

as differences between clusters are very significant, especially between the Non-Eco-Friendly and the Eco-Friendly. Looking at the bigger picture, it has to be said that clear-cut profiles are hard to draw based on this study's insights. Although some properties are identifiable, the clusters cannot be delimited from others in every attribute. Similarly, differences exist within clusters (e.g., income level for the Non-Eco-Friendly and level of education for the Eco-Friendly), illustrating the inconsistency that is symptomatic for analyses of consumers' attitudes and behaviors. Nevertheless, meaningful idiosyncrasies were identified in some parts, and summarized in Fig. 3.

3.3 Behavioral Outlook of Segments

After segmenting and profiling, a forecast for future behavioral tendencies of respective customer segments in air travel can be given, based on the behavioral segmentation variable.

3.3.1 The Non-Eco-Friendly

The Non-Eco-Friendly exhibit a relatively strong aversion to CO_2 ticket surcharges (concerning attitudes), and a comparatively low, but not as distinct desirability for substituting planes by trains. In fact, only 24% of this cluster state that they are unwilling to use trains instead of planes. A large stake of 40% is in the indecisive middle, not sure whether they will substitute planes for short-haul distances. This underpins the importance of strengthening the attractiveness of alternatives like trains, as this might be decisive for a hesitant subgroup like in this cluster. Another 36% expresses a modest desirability to effectively use trains for journeys within Europe.

3.3.2 The Attitude–Behavior Inconsistent

Remember the Attitude–Behavior Inconsistent cluster paradoxically shows the second highest pro-environmental score profile, but the lowest desirability for the use of trains. But since average values can be deceptive, also here the exact distributions must be considered. A third of the respondents are not willing to use trains as means of transport for short-haul distances. Another 26%, however, are undecisive, and a big proportion of 41% is considering using trains instead of planes. Also in this cluster, the first impression is misleading, as there is a substantial willingness to take on journeys by train.

3.3.3 The Moderate Eco-Friendly

The moderate Eco-Friendly cluster in contrast shows a clear picture when it comes to a possible use of trains. Although a few members of this cluster are against a CO_2 surcharge, the majority of respondents show pro-environmental attitudes. On the behavioral side, the desirability for the use of trains is evident, as none of the members of this cluster are against the use of trains, whereas one-fourth are in the insecure middle. A remarkable 76% show a clear desirability to take a train for journeys within Europe.

3.3.4 The Eco-Friendly

The last, smallest and most clearly distributed cluster is the Eco-Friendly one. As this cluster represents a pole on both, attitudinal and behavioral scale, 100% of respondents favor a substitution of planes by trains. This cluster therefore already sees trains as preferable means of transport when it comes to travel distances below 700 km.

Viewed holistically, the detailed analysis of behavioral tendencies of consumer segments shows a clear willingness to substitute planes by trains over all segments. Even the less environmentally friendly customer groups have fractions that support the idea of train travelling. Numerically spoken, only 17% of all apportioned respondents are not willing to use trains and prefer travelling by air. An impressive 57% are willing to resign from flying on short-haul distances and use trains instead. Another 26% of consumers form the undecisive middle, which have a medium desirability for train use, and are therefore susceptible for possible enhancements in railway infrastructure or increases in airfares. Furthermore, this latent acceptance for trains represents great potential for informative comparisons between train and plane journeys, as real travel times are more often shorter by train than consumers might expect. Important factors in this regard are the decision criteria, which influence consumers' choice for a specific mode of transport. This topic is dealt with in another study, whose insights are presented in the following section.

4 Decision Criteria for Modes of Transport

4.1 Critical Factors for Travel Mode Choices

The choice of mode of transport is a complex decision and has received considerable attention from academia. In the present context, the focus lies on product-related criteria that determine the modal choice of travelers. Previous academic work related thereto identified travel time, price, flexibility, safety, and familiarity with a specific travel mode, as the most central determinants for travel mode choice (Bieger & Laesser, 2004). Other studies added convenience and comfort in the context of air tourism and environmental concern (McDonald et al., 2015). In light of the emerging debate on climate change and aviation, sustainability criteria now play an increasingly important role in the decision-making process of consumers (Dällenbach, 2020). Therefore, based on previous findings in literature in the current context, price, travel time, and CO_2 emissions (as an approximation for sustainability) are defined as the critical factors in the decision-making process of transport means choices.

4.2 Influence of Knowledge on Travel Mode Choices

Building on these previously elaborated factors of travel mode choices, a study among Swiss students was conducted in order to ascertain the influence of knowledge of these factors on the choice of travel means. In this context, mode of transport choices, in accordance with previous chapters, were contained to trains and planes. As a tool to examine decision making processes and underlying importance of decision factors, the General Evaluability Theory by Hsee and Zhang (2010) was used.

> **General Evaluability Theory**
> According to the General Evaluability Theory (GET), decisions highly depend on how informed consumers are about products and attributes related to these decisions. Without adequate information, consumers are unable to make informed decisions. The GET postulates that the value sensitivity of consumers depends on the evaluability of these values. More precisely, with higher evaluability of attributes, value sensitivity increases. This means that consumers react stronger to changes in values to which they can relate. The evaluability thereby highly depends on knowledge gained through experience and is based on reference points gathered in the past.

Because of this importance of knowledge within the General Evaluability Theory, it is the primary focus of the following study design. In the survey among Swiss students, it was then tested how pronounced knowledge is about the three critical factors for transport means choice, and subsequently, if the choice of mode of transport differs when knowledge is manipulated (i.e., increased).

Results show that the knowledge of decision attributes among people differs. The results from the survey show that knowledge of travel time is greater than knowledge on prices, which in turn exceeds the knowledge on sustainability (or CO_2 emissions of the respective means of transport). However, the level of knowledge on these decision factors seems to play only a minor role in the decision-making process of travelers. Overall, the differences in knowledge are small and seem to have no significant effect on consumers' choices concerning the mode of transport.

Second, increasing knowledge about critical decision criteria does not influence choices regarding transport means. In two manipulations, subjects were told that (1) the actual travel time advantage of the plane over the train is smaller than widely believed, and that (2) travelling by plane is around 30 times more harmful to the climate than traveling by train (due to higher CO_2 emissions). The manipulations aimed at increasing knowledge about the two important decision factors travel time and CO_2 emissions, but the comparison with the control group showed that increasing the respondents' knowledge on these decision factors had no significant effect on their travel mode choices. The result, that increasing knowledge on CO_2 emissions does not affect travel mode choices, is in line with previous findings in which it was suggested that plane use may not be discouraged by increasing knowledge about the environmental impact of planes. In contrast, it is somewhat surprising that increasing knowledge on travel time too, had no impact on the travel mode choice, given that this is regarded as a promising strategy in encouraging people to travel by train.

Lastly, the respondents were more likely to take the train to Dusseldorf than to Berlin. They chose the more sustainable travel option (train) over the less sustainable plane on the shorter trip (Zurich—Dusseldorf), but preferred the plane on the long journey (Zurich—Berlin). Whereas train and plane tickets to Berlin cost the same in the choice experiment, the train ticket from Zurich to Dusseldorf was 20% more expensive than the plane ticket. This preference is remarkable because people did not care about the higher price of the train ticket to Dusseldorf, which suggests that sustainability and travel time are more critical for the choice of transport means than price—at least within a certain range. The importance of a certain range is another implication of results, as a critical threshold might exist in which trains are seen as a suitable alternative, albeit a higher corresponding price. This also entails that the willingness to substitute planes by trains decreases at a certain critical point in travel time.

Further analyzing open answers in the study reveals that three as many respondents chose the train over the plane in both scenarios, whereas a substantially smaller fraction consequently chose the plane for both constellations. This behavior might indicate that environmental consciousness plays a role in the means of transport choices and that some people tend to choose the train, no matter what. These findings thereby support the insights from the previous chapter on consumer segments, in which a broad acceptance for train use on short-haul distances, based on ecological motives, can be observed. The results of this study relativize the willingness to use the train insofar, as it is limited to a certain critical time and distance threshold, after which consumers are not willing to substitute trains by planes anymore.

5 Consumer Expectations Toward Sustainability

Derived from statements of the representative study and its open-ended question, expectations from customers in the context of flight shame and increased environmental consciousness are presented. These expectations are directed toward the industry and policy, but also to society and other consumers in general. The illustration shows the most common needs and demands expressed by customers in the representative study concerning future aviation, and are discussed in detail in the following (Fig. 4).

5.1 Changes in Consumer Behavior

The most common claim is directed not toward the industry, but to the consumers themselves. Generally, many respondents attach great importance to considerate action in the field of air travel. It is also evident that many people have a differentiated opinion when it comes to air travel. They are not particularly in favor or against it, they rather call for consciousness and rationality when flying or travelling in general. The demand for a conscious consumption is a popular demand from eco-friendly consumers, as well as from non-eco-friendly or skeptics. While the former use this argument to call for a reduction in air travel, the latter take it as a reason for self-responsibility and freedom of choice. However, a consensus of both groups can be found in the fact that that excessive air travelling, like for shopping trips abroad without any actual need, must be reduced.

A suiting example is the *Fly Responsibly* initiative of the Dutch airline KLM. Through the program, KLM seeks for a cooperative approach to obtain a more

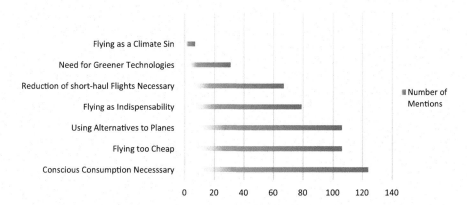

Fig. 4 Sustainability consumer expectations. Own illustration

sustainable aviation, inviting airlines, partners, customers, and employees. It highlights their own field of actions, e.g., use of sustainable fuel or closed loop recycling, and shows what already can be done from a customer perspective, like compensating CO_2 emissions or encouraging the use of alternative travel options. Further, it is showcased what the industry as a whole can and is doing to converge toward sustainable aviation. Always presented in a critical self-reflection, the program leaves an authentic impression of the airline's actions and efforts to make their business as environmentally compatible as possible and invites customers to join, without raising the admonishing finger.

5.2 Market and Policy Mechanisms

This goes in hand with the second most frequent consumer statement: Flying is viewed by many as a too cheap commodity, that needs to be increased in monetary value so that people re-begin to appreciate the real value of flying. Also, here not only supporters of flight shame hold this opinion, but also critics do. In this context, many respondents request airlines to resign from promoting very cheap flight offers, especially for short-haul flights. Anything else would run counter to the image of an airline as a responsible stakeholder in nowadays global climate issues.

The low price of flights was often mentioned together with the argument of using alternatives to planes. Many respondents indicated that they were already using trains for connections within Europe, but a considerable amount said they would do so too, if infrastructure and train connections would be better, especially for trips to Germany. In a similar vein, night trains were a popular request, that would cause many people to switch to trains as mode of transport. Relating thereto, a further limiting factor for switching is the higher price for train connections compared to prices for flights for the same route. For this aspect, many respondents called for political action to make railways competitive.

A further explicit necessity highlighted by consumers is the reduction of short-haul flights. Many are, as explained above, in favor of substituting planes by trains, with a supply-side reduction being requested. According to many consumers, a behavior change will not solely be initiated by customers, in fact, airlines must also induce change in this area by taking a proactive step. Interestingly, completely abstaining from flying was only stated by very few respondents. Also, harsh statements—like flying is a sin—were rare. Instead, a lot of consumers specified that the industry needs to develop technologies and alternatives like electric airplanes for a sustainable aviation, as air travel is still considered by many as indispensable part of tourism and travelling. Especially on longer, transcontinental journeys, no alternatives exist in the view of consumers.

6 Managerial Implications

The findings of the previous chapters can deliver valuable insights for industry representatives and policymakers. Derived from consumer perceptions of flight shame, the different customer segments built and their resulting expectations, the following implications can be drawn.

The discussion about flight shame is a controversial topic in Switzerland and Europe. Although opinions of people regarding flight shame are divided, there is a consensus on conscious air travel for the future. This includes a reduction of binge flying and numerous short-haul flights. In fact, the reduction of short-haul flights within Europe was an explicitly requested demand of consumers. Associated with this, prices for such flights are perceived as far too low and thereby creating nonexistent demand. In this field, airlines can make a proactive step and reduce the number of flights for very short distances. This also includes a change in marketing strategies, as flights advertised as bargains for shopping trips collide with the image of a reasonable and environmentally conscious airline. Furthermore, as conscious air travel is a widespread demand, it is recommended for airlines to initialize programs that address this topic by highlighting their own efforts in the field of sustainability, but also invite customers and other stakeholders to actively take their part in a transition toward more environmentally compatible air travel. As environmentally friendly attitudes exist in various consumer segments, but the step toward congruent behavior still remains intractable, airlines can incentivize this shift by offering bonus programs or reward schemes for customers who choose more environmentally friendly options within the service portfolio of an airline.

Related thereto, findings of this study revealed a majority of consumers willing to substitute planes for short-haul distances within Europe. It is even reinforced by the finding that for short-haul distances, consumers prefer the train option over the plane, despite the former being more expensive. This change in consumer behavior must be included in airlines' services creation. It involves not only an already mentioned reduction of short-haul flights, but also proactive steps to offer attractive products and seamless mobility. Similar ideas, like airlines offering train connections, have already been brought up by mobility researchers and politics. The insights of this study support the necessity of such new offerings, rethinking the role of airlines and their business model. Conceivable in this regard are cooperations with railway companies, that enable, e.g., a shared booking platform or mutual miles programs. Another keyword concerning this matter is intermodality. Different modes of transport have to be seen as complementary elements in an interconnected mobility system, maximizing comfort and accessibility for consumers. Partnerships and cooperations are therefore essential for airlines in order to offer not only a flight, but an entire journey.

For policymakers, these insights reveal two main scopes of action. Often discussed taxation of kerosene or a CO_2 surcharge are ways to increase prices and thereby internalize environmental costs. Of course, such measures are complex and difficult to implement, especially when multinational solutions have to be found to maintain competitiveness of national carriers. However, segment analysis has shown that two-thirds of Swiss consumers perceive a CO_2 surcharge as a desirable option, which therefore is worthwhile considering because it counteracts the not welcomed phenomenon of unreasonably cheap tickets. Another area where political action can be constructive is railway infrastructure. To foster and contribute to sustainable mobility, the quality of air travel alternatives is decisive. As a high willingness for the use of trains exists on the consumer side, good services will be well-received and are likely to be used quickly. This also requires intensified international cooperation between railways to meet consumer needs and make cities within Europe reachable with efficient and convenient connections.

Another key finding in this study is the differentiated and indecisive character of a large consumer base. Many individuals do not have a distinct opinion in favor of or against flying, which makes them susceptible to reasonable and persuasive arguments. One approach to benefit from this insight is promoting and highlighting a modern and state-of-the-art fleet, which can be a decisive argument for a consumer in Switzerland. This can also include data and facts for areas like fuel efficiency compared to older airplanes, the use of alternative fuels or general CO_2 reducing measures. Conceivable in this regard is also a collaboration with companies offering new technologies to capture CO_2 from the atmosphere and thereby going one step further than the already existing CO_2 compensation schemes. Such information could be integrated into on-board entertainment systems and presented appealingly to stimulate interest in the topic. Associated with this, a variety of consumers are requesting the development of alternative propulsion technologies, affirming that they accept and are willing to use them in the future. This should encourage airlines and suppliers to further enforce their engagement in developments of new technologies. These efforts should be communicated accordingly, in order to strengthen the image of a future-oriented and responsible airline.

7   Conclusion

At the beginning of this chapter, we have seen that Swiss consumers belong to frequent flyers globally. However, the growing environmental consciousness does not leave the area of air travel unaffected. Rising passenger numbers on intra-European railway connections indicate the first change in travel patterns. In this context, flight shame also exerts its influence on air travelers, as it is a very controversial topic among consumers. On the one hand, it enjoys a certain popularity and support, as 43% of consumers show positive attitudes toward the movement. On the other hand, 33% are clearly negative on the subject. Congruent with their attitudes, supporters of flight shame fly significantly less than adversaries of the movement. However, in the group with negative attitudes, typical denial patterns were observable, implying an existence of cognitive dissonance.

Further exploiting existing data yields a four-part consumer base, existing of the Non-Eco-Friendly, an Attitude–Behavior Inconsistent group, the Moderate Eco-Friendly, and the Eco-Friendly segment. Profiling these clusters shows significant differences in political party preference and the number of cars in households, while no particular differences were found in gender, age, education, and income. Looking at the consumer segments at a behavioral level, a high willingness to substitute planes by trains for short-haul routes was disclosed across all segments, as a majority of 57% displayed a high desirability for train use, while only 17% are strictly unwilling to substitute planes. Regarding the substitution of planes by trains, the analysis of related decision criteria for these modes of transport has yielded interesting results. An increase in knowledge of the critical decision criteria does surprisingly not have a substantial effect on substituting planes by trains. However, as long as the train journey is within a critical threshold in terms of distance and time, the willingness for using trains instead of planes overrules the price elasticity, meaning that the train is chosen over plane, although the train journey is more expensive. Expanding the focus of air travel behavior on long-haul distances, a reduction of plane used on these journeys can, however, not be expected, as long-haul flights are predominantly considered indispensable.

In the course of sustainability concerns, consumers expect airlines to reduce offers of ultra-cheap flights, as this collides with the image of a responsible contemporary airline. A further demand, which is more directed toward policymakers, is fostering the attractiveness of suitable alternatives like trains to enable a more environmentally friendly way of travelling. Strict measures like bans, however, are not considered appropriate, also because great importance is attached to self-responsibility and related conscious air travel.

References

Abend, L. (2019). The season of 'flight shame' takes off. *Time Magazine, 194*(6), 10–10.
Air Transport Action Group (ATAG). (2020). *Facts and figures*. Retrieved from https://www.atag.org/facts-figures.html
Bask, A., Halme, M., Kallio, M., & Kuula, M. (2013). Consumer preferences for sustainability and their impact on supply chain management: The case of mobile phones. *International Journal of Physical Distribution & Logistics Management, 43*(5/6), 380–406. https://doi.org/10.1108/IJPDLM-03-2012-0081
Bieger, T., & Laesser, C. (2004, March 25). *The market entry of low cost airlines (LCA): Implications for mode choice between Switzerland and Germany*. 4th Swiss Transport Research Conference (STRC), Monte Verita, Asona. Retrieved from http://www.alexandria.unisg.ch/publications/19271
Credit Suisse. (2019). *Credit Suisse Sorgenbarometer 2019*.
Dällenbach, N. (2020). Low-carbon travel mode choices: The role of time perceptions and familiarity. *Transportation Research, 86*. https://doi.org/10.1016/j.trd.2020.102378
Ehrbar, S. (2020, March 5). *Von wegen Flugscham: Jeden Tag fliegen 3500 Personen nach London-Heathrow*. Tagblatt. Retrieved from https://www.badenertagblatt.ch/wirtschaft/von-wegen-flugscham-jeden-tag-fliegen-3500-personen-nach-london-heathrow-136572523
Gerretsen, I. (2019, December 17). *Thousands of people have stopped flying because of climate change*. CNN Travel. Retrieved from https://edition.cnn.com/travel/article/train-travel-flying-climate-scn-intl-c2e/index.html
Gössling, S., Humpe, A., & Bausch, T. (2020). Does 'flight shame' affect social norms? Changing perspectives on the desirability of air travel in Germany. *Journal of Cleaner Production, 266*. https://doi.org/10.1016/j.jclepro.2020.122015

Graham, A., & Metz, D. (2017). Limits to air travel growth: The case of infrequent flyers. *Journal of Air Transport Management, 62*, 109–120.

Henley, J. (2019). #stayontheground: Swedes turn to trains amid climate 'flight shame'. *The Guardian.* Retrieved from https://www.theguardian.com/world/2019/jun/04/stayontheground-swedes-turn-to-trains-amid-climate-flight-shame

Hsee, C. K., & Zhang, J. (2010). General evaluability theory. *Perspectives on Psychological Science: A Journal of the Association for Psychological Science, 5*(4), 343–355.

Institute for Economy and the Environment (IWÖ-HSG). (2020). *10th Consumer Barometer of Renewable Energy.* Retrieved from https://iwoe.unisg.ch/wp-content/uploads/Technical-report_Consumer-Barometer_2020.pdf

Intraplan. (2019). *Monitoring der Wettbewerbsfähigkeit des Schweizer Luftverkehrs 2018.* Retrieved from https://www.admin.ch/gov/de/start/dokumentation/medienmitteilungen.msg-id-74943.html

Juvan, E., & Dolnicar, S. (2014). The attitude-behaviour gap in sustainable tourism. *Annals of Tourism Research, 48,* 76–95. https://doi.org/10.1016/j.annals.2014.05.012

Kollmuss, A., & Agyeman, J. (2002). Mind the gap: Why do people act environmentally and what are the barriers to pro-environmental behavior? *Environmental Education Research, 8*(3), 49–58. https://doi.org/10.1080/1350462022014540

Kroesen, M. (2013). Exploring people's viewpoints on air travel and climate change: Understanding inconsistencies. *Journal of Sustainable Tourism, 21*(2), 271–290. https://doi.org/10.1080/09669582.2012.692686

McDonald, S., Oates, C. J., Thyne, M., Timmis, A. J., & Carlile, C. (2015). Flying in the face of environmental concern: Why green consumers continue to Fly. *Journal of Marketing Management, 31*(13–14), 1503–1528.

Meier, J. (2019, October 19). Dank Flugscham: Die Eisenbahn erlebt eine Renaissance. *NZZ am Sonntag.* Retrieved from https://nzzas.nzz.ch/wirtschaft/dank-flugscham-die-eisenbahn-erlebt-eine-renaissance-ld.1516511

Patel, T. (2020, February 13). *What is flying shame? Is it a movement with legs? Bloomberg Green.* Retrieved from https://www.bloomberg.com/news/articles/2020-02-13/what-is-flying-shame-is-it-a-movement-with-legs-quicktake

Stalder, H. (2017, May 16). *Jeder Schweizer fliegt im Jahr fast 9000 Kilometer.* Neue Zürcher Zeitung.

The Swiss Parliament. (2019). *Fertig mit der Schönfärberei. Wie gross sind die Wirkungen des Flugverkehrs auf das Klima wirklich* [Interpellation]? Retrieved from https://www.parlament.ch/de/ratsbetrieb/suche-curia-vista/geschaeft?AffairId=20194281

Wittmer, A., Puls, R. (2019). *Civil aviation 2040 – A perspective of the Swiss Society.* Center for Aviation Competence, University of St. Gallen.

WWF. (2019, June 17). *Schweizer bleiben am Boden fürs Klima [Press release].* WWF. Retrieved from https://www.wwf.ch/de/medien/schweizer-bleiben-am-boden-fuers-klima

Intermezzo: Considerations on the Interdependence of Technology, Consumer Behaviour Change and Policy Interventions to Achieve Sustainable Aviation

Alexander Stauch

Abstract

- This chapter serves as an intermediate discussion on the role of technology, policy, and behaviour change to achieve sustainable aviation and is intended to serve as preparation for the following chapters.
- To have a better overall understanding, future scenarios from a techno-optimistic point of view are contrasted with scenarios based on political interventions and behavioural changes.
- Based on a combination of techno-optimistic and behavioural change-based scenarios, a path towards sustainable aviation is outlined.

This short intermezzo discussion looks into the future and outlines linkages between scenarios based on technological progress and demand, and policy options. As the name suggests, this chapter is meant as an interlude to set the stage for the following chapters, based on the technology and customer sections from the previous chapters. It especially addresses the question of whether the airline industry should rely solely on technological progress as a solution to the GHG emissions problem, or whether there are also other levers that can be used to achieve long-term structural change in the industry towards sustainability. To do so, the discussion starts with a general outlook on the future development of the aviation industry. This will be followed by a discussion regarding the influence of different political measures on demand, and what the aviation industry could learn from other industries that have already experienced strong political interventions. Then, the discussion continues by linking the different scenarios to specific policy measures but also other risks (such as fuel price shocks) that could shape future aviation demand. Additionally, the scenarios

A. Stauch (✉)
Institute for Economy and the Environment, University of St. Gallen, St. Gallen, Switzerland
e-mail: alexander.stauch@unisg.ch

will be plotted using two different dimensions, demand growth on the x-axis and GHG reduction based on the penetration of new and climate-friendly aviation technologies on the y-axis. After we know which pathways (scenarios) will lead to which outcomes in terms of demand growth and GHG reductions, we will continue the discussion about which pathways should for which reasons be desired to really achieve sustainable aviation in the long run. These pathways not only include political measures, but also technological progress and behavioural change of consumers.

1 Scenarios: General Outlook on Aviation GHG Emissions and Demand Until 2050

Scenario planning and analysis is a strategic management tool, often used to analyse and to picture possible future developments and to present them in a coherent way. This method also allows the inclusion of factors that are difficult to formalise, such as novel insights about the future, deep shifts in values and behaviour, but also unprecedented regulations or inventions. In this chapter, we use the four scenarios presented by the Air Transport Action Group[1] (ATAG, 2020) as the outlook for the aviation industry until 2050. The central traffic forecast used by ATAG shows that by 2050, approximately 10 billion passengers per year will travel a distance of 20 trillion revenue passenger kilometres, which is around three times more than 2019. Without interventions (maintaining the current fleet and current operational efficiencies), this activity would generate about 1800 megatons of CO_2 (compared to 640 megatons CO_2 in 2019) and require over 570 megatons of fuel per year. This baseline (or business as usual) scenario would lead to a situation where the climate targets are completely out of reach.

The following four ATAG scenarios outline how the industry uses technology, operations, infrastructure, offsets, and sustainable aviation fuels to reduce this value, despite the high growth levels in demand, to meet the industry climate target of 325 megatons CO_2 emissions per year by 2050 and exceed it in subsequent years. Therefore, these scenarios are categorised as "technology-driven scenarios".

1.1 Technology-Driven Scenarios

Baseline Scenario (1): The baseline scenario is a continuation of current efficiency trends in all pillars of action, with no acceleration of improvements in other sustainable aviation technologies. Technology improvements are conservative (i.e.,

---

[1] ATAG provides a platform for the commercial aviation sector to work together on long-term sustainability issues. It is funded by its members. These include airports, airlines, airframe and engine manufacturers, air navigation service providers, leasing companies, airline pilot and air traffic controller unions, aviation associations, chambers of commerce, tourism and trade partners, ground transportation and communications providers.

assuming no to little risk by shifting to unconventional platforms) and therefore show a continuation of the current rate of improvement, with another wave of new aircraft joining and starting to replace the fleet around 2030–2035. Sustainable aviation fuel is developed and introduced based on current rates resulting in 4% to max 25% GHG reduction in 2050, while 50% to 70% of the emissions remain in this scenario. This would require high offsetting or other technologies, such as carbon capture, to reach net-zero emissions by 2050.

SAF Scenario (2): This scenario assumes that there is not a significant shift to electric or hybrid aircraft, with the industry prioritising heavy investment in sustainable fuels. SAF will reduce GHG emissions up to 75% until 2050, while technology, infrastructure and operations efficiency reduces the remaining 25% of the GHG emissions. This means that in 2050, net zero GHG emissions are achieved without relying on carbon offsets or carbon capture. Offsets are not expected to play a central role in meeting the 2050 goal but may be relied on between 2035 and 2050 as a transition mechanism.

SAF and Technology Scenario (3): Under this scenario, technology improvements are prioritised and ambitious with the expectation of the emergence of unconventional airframes and a transition of the fleet towards hybrid/electric aircraft from 2035/2040, starting with short haul flights. Significant investments in operations and infrastructure improvements result in substantial improvements and GHG emission reductions. The gap between GHG emissions after technology, operations and infrastructure improvements and the 2050 carbon goal is fulfilled by the use of sustainable aviation fuels, which will still account for around 65% of the GHG emission reduction by 2050, especially for long-distance flights. Similarly, as in SAF Scenario (2), offsets are not expected to play a central role in meeting the 2050 goal but may be relied on during 2035 and 2050 as a transition mechanism.

Technology Scenario (4): In this scenario, technology improvements are very ambitious with electric aircraft and hydrogen aircraft (powered by green hydrogen) as zero emission aviation technologies are developed for the 100–200 seat segment and hybrid-electric (fuel cell or battery combined with SAF) powered aircraft configuration for larger aircraft. SAF will account for around 50% of the GHG emission reduction by 2050, while new aircraft technologies using battery or hydrogen account for around 40% of the GHG emission reduction, while the remaining 10% of GHG emissions will be reduced based on efficiency and operational improvements.

In all scenarios, SAF seems to play a major role, either for the transition towards net zero GHG emissions using new aviation technologies (battery or hydrogen) or even as the ultimate hope to fulfil emission targets by 2050 entirely by using SAF. Since SAF is already technically feasible today and only requires few adjustments to aircraft technologies and ground infrastructure, a strong development of SAF can indeed be expected in the medium term. However, the aviation industry should not forget to focus on new and emission-free technologies and to push forward the transformation in this area. Too much penetration with SAF could lead to the industry increasingly focusing on SAF only, as it is cheaper and involves less transformation costs and risks. This could slow down or even prevent the development of truly new technologies (battery and hydrogen) in the long term. Other scenarios provided by different stakeholders in the aviation industry are in general quite similar to the four scenarios from ATAG. The scenarios usually vary in terms of SAF, new technology (electric, hybrid, hydrogen) penetration and how large the respective shares of new technology and SAF could be until 2050. Thus, the four scenarios from ATAG represent a good summary of several scenarios that are currently discussed within the industry.

What all the four scenarios, including most of the other scenarios coming from the industry, have in common is that they assume a steady growth of aviation demand of 3% per year, resulting in an overall aviation demand level that will be around 2,5 to 3 times higher in 2050 compared to 2019. Assuming, however, that there will also be political interventions and/or changes in consumer behaviour in the future, e.g., in the form of increased CO_2 prices or increased climate awareness, the average annual growth in demand of 3% could also move to a lower level, or even become negative. Increased prices and consumer behaviour changes can have a high influence on the demand, depending on the price increase and the elasticity of the demand curve. Thus, we will take a closer look at prior political price interventions and its effects on demand in order to understand how future aviation scenarios based on price interventions and realigned consumer behaviour could look like.

1.2 About Policies and Demand

The future development of industries has often been influenced by political measures. To date, however, more demand-stimulating measures have been taken for the international airline industry, as the development of international air travel has been seen by policymakers as a sign of progress and global connectivity. These stimulus measures include, for example, tax exemptions on kerosene. In the meantime, however, the political discussion is moving in a different direction for the airline industry as well. In the wake of the climate targets, the airline industry as a beneficiary of outdated policy measures is increasingly in the focus of politicians, who are now thinking more about climate-political measures for the airline industry. However, due to a lack of experience of how these measures could look like and how strong they will be, it is difficult to assess how climate policy measures will affect future demand in the aviation industry. Of course, it always depends on how drastic the measures are as well. Therefore, we would like to take a brief look at another industry that was also affected by a variety of different and strong regulatory measures: the tobacco industry. At this point, we would like to point out, that we are not directly comparing aviation with the tobacco industry, but with the fossil fuel industry as an important resource provider for aviation. As with fossil fuels, global tobacco companies had a very strong lobby that was able to prevent these heavy regulations for a long time. However, the growing body of data and scientific findings increasingly confirmed that smoking is very harmful to the human body. The same can be said of CO_2 and our earth's climate. Here, as well, the data and the scientific findings are becoming more and more concentrated, underlining that the

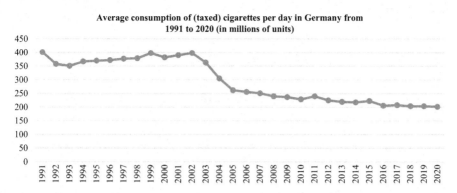

Fig. 1 Demand decline of cigarettes in Germany from 1991 to 2020. Source: Statista (2021)

burning of fossil fuels is significantly driving man-made climate change. Based on these considerations, it is only a matter of time before political decision-makers enact drastic measures to reduce the use of fossil fuels in all sectors (not only, but also in aviation). But how could these measures look like and how effective will they be in reducing demand? Again, we try to learn something from the tobacco industry, which has already experienced a lot of different regulations. Looking at historic data, the demand for tobacco products in Western-European countries has declined significantly since the increasing political regulations from 2002 onward, as the following graph illustrates based on the cigarette consumption in Germany as an example.

If one takes a closer look at Fig. 1, one could assume that strong price increases due to increased taxes on tobacco products were mainly responsible for the sharp decline. But in fact, using the example of Germany, it was a combination of different measures, with price being one of the factors that had an influence. In addition, much has been achieved through raising awareness about the harmfulness of cigarettes. In 2002, for example, the European Union decided that all cigarette packages had to carry warnings (e.g., smoking ages the skin, makes people infertile and can lead to a "slow and painful" death), from 2004 these warnings also included pictures of body damage caused by smoking. In 2003, cigarette advertising in print media and at sporting events was also banned by the EU regulators. Consequently, a mix of a wide variety of measures has led to a sharp drop in demand, especially from 2002 to 2004. In addition, since 2002, taxes on tobacco products have been regularly increased, so that the price of cigarettes is constantly rising. However, this price increase has only led to a limited decline in demand in subsequent years. Other countries, such as England or New Zealand place much higher price increases on cigarettes, which in turn led to a much stronger decline in demand than in Germany. But what does this all mean for the future of the aviation industry? First of all, it means that regulations can have a strong impact on the demand, even if the industry had a strong lobby until a certain point in time. Secondly, it means that measures only tackling the price, such as taxes or carbon pricing, might not be sufficient to reach the desired outcome when the measures are not designed ambitiously enough. Thirdly, the combination of awareness and education about the problem in general will also be key. This means, that if governments start to act, they should also consider a communication plan that explains very well why the measures have been taken and which goals are to be achieved with them. Furthermore, the government should also promote alternatives that could replace the customer need with new or alternative technologies.

Looking back in time, where aviation prices were much higher in general, it can be seen that the enormous growth in demand for air travel is primarily due to the sharp drop in prices and the emergence of low-cost airlines. Conversely, this would mean that an increase in prices could weaken this strong growth in demand. However, a lower growth in demand may also have other reasons besides political interventions. With increasing awareness of the climate problem, many consumers are realising that excessive air travel is very harmful to the climate (keyword "flight shame"). In addition, other external shocks such as global pandemics, war zones (e.g., East-Ukraine or Syria), increasing number of environmental disasters (volcanic eruptions like in Iceland, desert storms, hurricanes, etc.) and energy price shocks can have a strong impact on air travel demand as well. Therefore, the introduced scenarios will be enlarged with the dimension of external demand shocks, which need to be considered as well. Thus, we additionally picture four generic scenarios, that are mainly shaped by political interventions but also by other external shocks that could affect demand in the long run. Therefore, these scenarios are categorised as "policy driven scenarios".

1.3 Policy-Driven Scenarios

Baseline Scenario (1): The baseline or business as usual scenario assumes no major price shocks or political interventions that would affect air travel prices substantially. Also, additional external shocks are not assumed, which will stabilise the growing demand for aviation while no major achievements in the large-scale deployment of sustainable aviation technologies (including SAF) are made.

Technology Incentive Scenario (2): In this scenario, it is assumed that political actors implement an incentive structure to develop new sustainable aviation technologies. These incentives can include governmental funds for research and direct or indirect subsidies for suppliers of sustainable aviation technologies. At the same time, no pricing or other penalty measures for the use of GHG emitting technologies have been implemented by policies.

Price Shock Scenario (3): This scenario assumes a price shock on air travel resulting either from political interventions on GHG emission pricing (such as a CO_2 tax or emissions trading) or from other external fuel price shocks (shortage in oil supply, political instability in oil-producing countries). At the same time, no incentives for new and sustainable aviation technologies are in place, which means that no alternatives for air travel are in place, thus resulting in lower or even negative demand growth.

Price Shock and Technology Incentive Scenario (4): This scenario assumes that a price shock on GHG emitting technologies, based on political interventions or external fuel price shocks combined with an incentive structure for new and sustainable aviation technologies, will also reduce demand growth. At the same time, it will lead to a situation from which demand can return to normal growth in the future. First, the price shock on GHG emitting technologies will reduce demand and lead to a situation where sustainable aviation technologies are superior in terms of prices compared to conventional aviation technologies. With lower levels of demand, the transition of the aviation system, including the primary energy infrastructure, will be less complex and less capital intensive at the beginning. Once the transformation is achieved, efficiency gains will reduce prices and growth levels could come back in the far future (after 2060).

It is important to state at this point that the four generic scenarios only present external shock scenarios that will affect the demand side in the long run. The aviation industry scenarios from ATAG should be compared and in the best case combined with the generic demand shock scenarios to really have a holistic view on the future of the aviation industry. In the following paragraphs, we will discuss different aspects of these scenarios in more detail and then classify and combine these scenarios to form an argumentation line towards a scenario that will lead to real sustainable aviation.

1.4 The Link Between Scenarios and Policies

The[2] scenarios dealing with the future development of the aviation industry presented at the beginning of this chapter have demonstrated that a wide range of different futures is possible, differing in the development of global demand and the primary energy source for aviation by 2050. Also, the aviation industry scenarios from ATAG should be compared and, in the best case, combined with the generic demand shock scenarios to really have a holistic view on the future of aviation industry. Conceptually, the scenarios can be distinguished into the four quadrants presented in Fig. 2. The left part of the graph represents a high-demand growth future for the aviation system, while the right part of the graph is characterised by lower or even stagnating demand growth. The lower part of the graph shows scenarios with a low penetration of new and climate-friendly aviation technologies[3] that reduce GHG

---

[2] The following sections of this intermezzo chapter have been inspired by Wüstenhagen (2013); further references used to build the argumentation include Gardezi & Arbuckle (2020), Hankey & Marshall (2010), Kirby & O'Mahony (2017), Sanden & Azar (2005), Spence & Pidgeon (2009), Teske et al. (2011), Tran (2016), Trutnevyte et al. (2019), Unruh (2002), Von Weizsäcker et al. (1998), Voß et al. (2009), West et al. (2010) and Williamson et al. (2018).

[3] New and climate-friendly aviation technologies involve hybrid (SAF and battery), hydrogen and all-electric aircraft. The use of SAF only is not considered as fully climate-friendly technology, since it still produces a lot of GHG emissions during operation at an altitude, where climate gases tend to have an even greater effect than at ground level due to radiative forcing (change in the earth's energy balance due to change in the effect of radiation from space, caused by, e.g., aviation induced cloudiness from SAF emissions (see also Sect. 2.1 or Lee et al., 2021)) and thus, despite its carbon-neutrality, still has an impact on global warming.

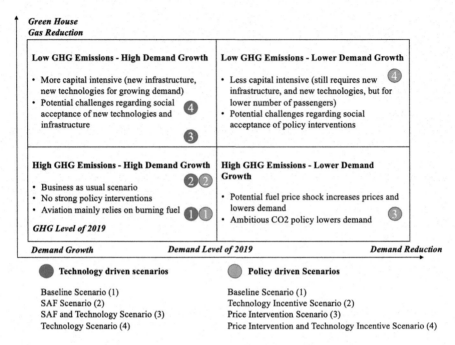

Fig. 2 Scenarios for high- and low- GHG emissions combined with high and low demand growth in the aviation industry. Own Illustration

emissions, while the upper part of the graph represents future aviation systems with a high penetration of new and climate-friendly aviation technologies.

The baseline (or business as usual) scenarios are placed in the lower left quadrant of the graph, which would combine a low share of GHG-reducing technologies with high levels of demand growth. Scenarios in this quadrant tend to result in higher GHG emissions compared to 2019. These high-demand growth, low climate-friendly technology penetration scenarios would not only make the world vulnerable to climate risk but also imply significant policy intervention and fuel price risk for operators.

The lower right quadrant of Fig. 2 depicts a world with relatively low penetration of new and climate-friendly aviation technologies in combination with relatively low demand growth. Such scenarios could be a result of a shortage of fuel (oil) supply or strong changes in oil demand—either because of fuel price shocks or of active policies—while at the same time no particular efforts were made to support new and climate-friendly aviation technologies, indicating that there may still be a lot of challenges with regard to climate stabilisation and the transformation towards sustainable aviation.

The upper half of the graph depicts two types of energy futures with widespread deployment of new and climate-friendly aviation technologies ("Low GHG scenarios"). The scenarios in the upper left quadrant combine significant expansion of new and climate-friendly aviation technologies with strong growth in demand and supply, whereas the upper right quadrant represents a combination of a high share of new and climate-friendly aviation technologies with lower levels of demand growth, to the extent that some of the most extreme scenarios assume that global demand could only be a bit higher in 2050 compared to the base year 2019. When comparing these two versions of a high penetration of new and climate-friendly aviation technologies world, a few differences become apparent. First, while the scenarios in the upper right quadrant tend to reach low levels of GHG emission concentration, there is more variation in the upper left quadrant. In other words: a high penetration of new and climate-friendly aviation technologies, combined with high-demand growth future, seems to bear a higher risk of overshooting global carbon targets. Second, given the need to create a sustainable aviation technology system at a large scale in a world characterised by high demand, the scenarios in the upper left quadrant are arguably more capital intensive than the ones leading to the upper right quadrant. Transforming a strong growing industry compared to an industry with stable growth leads to lower transition time, less investments, and greater opportunities to achieve growth in demand after the transition.

Finally, there are different societal risks involved in the two kinds of high penetration of new and climate-friendly aviation technologies scenarios. Those scenarios that combine high penetration of new and climate-friendly aviation technologies with low demand growth rely on either active demand reduction policies or they assume significant fuel price shocks, both of which may create barriers to political and social acceptance. On the other hand, the high penetration of new and climate-friendly aviation technologies with high demand growth scenarios rely on greater levels of deployment of new and climate-friendly aviation technologies and its supply infrastructure, which in turn could also be a social acceptance issue in many countries (e.g., developing renewable energy projects, such as wind or solar farms to produce green electricity or green hydrogen for aircraft propulsion).

As outlined, Fig. 2 summarises the scenarios and plots them into the four described quadrants characterised by GHG reduction (industry-wide) and demand growth reduction. Most of the scenarios are placed in the lower left quadrant, assuming a continuation of high emissions and high demand growth. The SAF Scenario (2) is also placed in this quadrant, since it assumes no major changes in aircraft technologies (e.g., hydrogen or battery). The main concern with using SAF only is that it still produces a lot of GHG emissions during operation, especially at an altitude where climate gases tend to have an even greater effect than at ground level. In combination with high growth, this scenario will most likely not decrease GHG emissions compared to 2019. Additionally, the widespread use of SAF could slow down or even prevent the development of new sustainable aviation technologies. Nevertheless, since SAF is already an improvement in terms of GHG reduction, compared to the baseline scenarios where conventional fuel still plays a major role. The Technology Incentive Scenario (2) is also placed in the lower left part. Technological incentives alone will not change much in the airline industry, as the entire industry is attuned to fuel consumption and has already perfected this to a very high level. Without a clear and effective pricing system for CO_2, incentives for new technologies alone would only lead to minor changes, such as the increased use of SAF. On the other hand, pricing systems for CO_2 alone do not necessarily lead to the desired change either. That is why the Price Incentive Scenario (3) is placed in the lower right quadrant. It may reduce demand to a certain level (depending on the size of price intervention and elasticity of demand), but it does not necessarily make supply more environmentally friendly. What is needed is a policy framework and a roadmap, i.e., a mix of incentives for new technologies on the one hand and a mechanism to curb demand for emissions-intensive travel on the other hand. This mix is represented by the Price Intervention and Technology Incentive Scenario (4) in the upper right part of Fig. 2.

The SAF and Technology Scenario (3) is placed on the border to the upper left quadrant. This classification is based on the fact that in this scenario, part of the air traffic (i.e., short-haul flights) will actually be carried out with new technologies by 2050, while all other flights will be carried out using SAF. However, since a strong growth in demand combined with a large share of SAF is still assumed, significant GHG emissions are still very likely in this scenario. The increasing demand also entails the risk that it cannot be completely met with the new technologies and SAF, and that the transformation could therefore be prolonged significantly, which also means a prolonged reduction of GHG emissions. The Technology Scenario (4) assumes that a large share (up to 50%, short and medium distance flights) of the aviation industry is already transformed by very aggressive deployment of new sustainable aviation technologies including hybrid-electric, electric and hydrogen aircraft. Only long distance will be covered using SAF by 2050. This scenario offers the greatest potential in GHG reduction without reducing demand, but also implies very high investments in new technologies and energy infrastructure. Additionally, this scenario bears the risk to hope for future technologies that do not yet exist in the market. A delay in technology developments resulting in a prolonged transition time of the industry bears several climate risks, especially when demand is still growing.

The placement of the scenarios in Fig. 2 can, of course, be debated. However, it can be concluded that scenarios that rely only on technological progress and assume constant growth in demand at the same time entail various risks. These risks include the possibility of slower technological progress than assumed and thus the danger of a very long or delayed transformation phase, which would even be reinforced by the steady growth in demand, putting the achievement of the climate targets in serious danger. In addition, it takes much more investment to transform a growing industry than one with stagnant or declining growth.

2 A Sustainable Industry Transformation Results from a Combination of Innovation and Behaviour Change

An important, yet often implicit dimension of future aviation scenarios is whether they assume changes to be mainly driven by technological Innovations, or whether they assume changes in behaviour as a driver for future development of aviation and

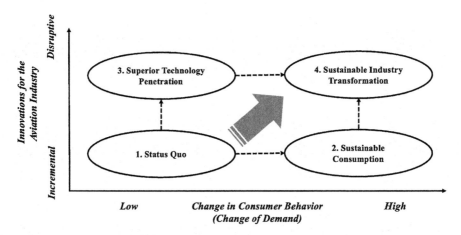

Fig. 3 Aviation innovations and behaviour change for a sustainable industry transformation. Own Illustration

mobility systems. Figure 3 outlines the spectrum of possible futures according to these two dimensions. The vertical axis distinguishes futures characterised by incremental technological innovations from those based on disruptive technological innovations. The horizontal axis characterises the level of behaviour change that is implied (incremental or disruptive). Most of the scenarios from Fig. 2 assume that both technological development and behaviour change remain incremental, leading to a continued rise in aviation demand and a focus on the deployment of existing aviation technologies (internal combustion engine, in the best case with increased use of SAF). Therefore, a high penetration of new and sustainable aviation technologies is relatively unlikely, which supports to maintain the "Status Quo (1)". In a behaviour optimistic world of the future, scholars argue that a shift in consumer behaviour will lead to a strong increase in "Sustainable Consumption (2)". Scenarios in this cluster assume that due to changes in lifestyles (e.g., climate youth and/or flight shame) and increased awareness for energy and climate issues, aviation demand decreases while the penetration with existing aviation technologies and SAF continues. Such disruptive behaviour change may be supported by active policies, e.g., aviation GHG emission disclosure and awareness campaigns, and could also be a result of external constraints, such as fuel price shocks. In a techno-optimistic[4] world, it is assumed that "Superior Technology (3)" solutions are the main drivers to solve environmental and societal problems while people can stick to their usual consumption patterns (e.g., instead of flying using kerosene, people will fly using hydrogen). Almost all scenarios on the future of air transport that come from the aviation industry itself, including the ATAG scenarios presented in the beginning,

---

[4]Techno-optimism is a "belief in human technological abilities to solve problems of unsustainability while minimizing or denying the need for large-scale social, economic and political transformation" (Barry, 2016, p. 3).

have a tendency for a rather techno-optimistic worldview. Scenarios in this camp tend to assume that disruptive technological change (e.g., availability of hybrid, all-electric and hydrogen aircraft) may result in a quantum leap in the competitiveness of sustainable aviation technologies, leading to higher market penetration (based on similar arguments related to other examples of low-carbon technologies). However, behaviours and lifestyles are still based on the status quo consumption, leading to efficiency improvements that may be compensated by growing demand and rebound effects (Dimitropoulos (2007)), and hence levels of aviation demand remain high. An analysis from Alexander and Rutherford (2019) indicates that technological efficiency gains in the energy sector were compensated through rebound effects based on higher demand afterwards, resulting in an even higher energy and resource consumption, or at least did not lead to a reduction. This was also the case for aviation in the past. Since 1990, GHG emissions per passenger kilometre have decreased over 50% due to efficiency gains in technology (ATAG, 2020), but the absolute GHG emissions from the aviation industry still increased over 200% since 1990 (Lee et al., 2021). Further, a successful market diffusion of innovative technologies usually needs some time, especially when a whole ecosystem is affected by the transition, and thus, high consumption levels would extremely slow down the transition process towards sustainable aviation. Additionally, putting all hopes on superior technology that is yet to be invented bears the risk of technological penetration delay, which would lead to a high likelihood of overshooting carbon targets. In conclusion, it can be said that relying solely on "Superior Technology Penetration (3)" entails too many risks of failing to meet the climate targets. If policymakers would like to build their climate change mitigation policies on an aviation system with a high penetration of new and sustainable aviation technologies, it is likely that disruptive changes will have to occur on both levels. The following Fig. 3 illustrates and summarises the way towards a "Sustainable Industry Transformation (4)", based on consumer behaviour change and sustainable aviation technologies.

2.1 Policy Options for an Accelerated Transition Towards a Sustainable Aviation Industry

Achieving a high level of new and sustainable aviation technology penetration combined with change in consumer behaviour represents a full industry transformation from today's aviation systems. Facilitating such a transition based on technology and behavioural change will require active policy approaches for the following reasons:

- A stable and predictable policy framework as well as clearly communicated long-term targets are required to attract long-term sustainable investments; otherwise investors will shy away from such investments due to the perceived policy risk.

- The required infrastructure investment (clean energy production, storage and distribution, but also low-carbon travel alternatives) will require some level of public funding or public-private partnerships
- While low levels of GHG reduction from aviation can be achieved with a relatively limited number of new technologies (e.g., SAF), a sustainable aviation world is likely to rely on a broader portfolio of new and sustainable aviation technologies with differing levels of maturity. Sustained efforts in research, development and deployment at significantly higher levels than today will be required to bring these different technologies to market over time.
- Technology R&D alone is likely not going to be sufficient to ensure commercialisation of new energy technologies. There will be a need for specific deployment policies to create protected spaces for experimentation with new and sustainable aviation technologies, and a subsequent scale-up of promising concepts. Additionally, there will be a need for government programmes (e.g., Rahm, 1993; Lerner, 1999) that aim at facilitating technology transfer to private firms, and private investors, especially venture capitalists, working towards the same aim.
- While some scenarios assume high penetration of new and sustainable aviation technologies at relatively high levels of demand, and technical potential is high for sustainable aviation, a superior technology & high-demand scenario tends to face tighter constraints when it comes to capital requirements and transition speed issues than a superior technology scenario that simultaneously decreases demand. Such a demand decrease may be driven by market forces (e.g., fuel price shocks), lifestyle choices (climate youth, flight shame) or by active policies (e.g., carbon pricing, energy and flight taxes, efficiency standards, labelling).
- Both, the level of demand and the share of sustainable aviation technologies in the mix depend on strategic choices made today that are heavily interconnected to other policy areas, notably energy infrastructure (generation, storage and distribution) planning and other mobility policies. Achieving a sustainable aviation world will depend on early policy integration.
- The magnitude of changes will require public consent to a variety of policies, which in turn implies increased efforts to raise public awareness for sustainable aviation and transport.
- To facilitate a change in today's travel behaviour, there need to be sufficient low-carbon alternatives to commercial flying: e.g., investment in high-speed train networks, which will require international and political coordination as well as funding.

2.2 Aviation Industry Should React Proactively and Prepare for Industry Transformation

If climate policy will be taken seriously, it is very likely that many policy interventions to reduce GHG from aviation will take place in the future. As an industry, there are different ways how one can react to such an outlook. A passive or denying view on this outlook could lead to the assumption that such political measures or strong changes in demand due to lifestyle changes can be absorbed by counter-campaigns, marketing and good lobbying. However, the most successful lobby of all times, that of fossil energies, is already encountering increased political resistance. Another possible reaction to this outlook could be to put all hopes for sustainable aviation on future technologies, implying that the industry simply has to wait until they are ready to solve the whole problem. This way of thinking is rather short-sighted and additionally, as outlined before, it bears a lot of different climate and financial risks as well. The past has shown that technology alone cannot solve environmental problems when demand growth remains high, as all technological gains for the environment are compensated by the increased demand. As the aviation industry expects a growth between 250 and 300 percent until 2050, it is thus highly questionable if progress in technology alone can compensate all the new emissions or even lower the emission levels compared to today. Consequently, we argue in this chapter that a proactive way to achieve the climate goals includes self-regulatory industry measures and a general paradigm shift in pricing and demand growth. The industry could, for example, set itself ambitious targets and at the same time integrate the emissions caused into the pricing mechanism. This would be equivalent to a political price intervention, with the advantage that the intervention would have the backing and consensus of the industry. At the same time, the industry should think about who will take on what role and what investments for the transformation, and what the industry should look like in 2050 to be fully sustainable. A clear vision for sustainable aviation 2050, backed by the entire industry and not by purely relying on technology, will be crucial for a successful transformation. Of course, finding a global consensus that increases prices and decreases demand is always difficult and maybe not very likely to happen. Nevertheless, if the industry cannot regulate itself, or if there is no global consensus within the industry for the pricing of GHG emissions, at least proactive industry actors should do so already today, step by step, to protect themselves against future interventions, which might result in future competitive advantages over competitors who did not prepare for the sustainable transformation.

3 Conclusion: Sustainability Requires Technology Paired with Lower Demand Growth, Proactive Airlines and a Reliable Policy Framework

New and sustainable aviation technology is one of the options policymakers have at hand to mitigate GHG emissions from aviation, while policy interventions for pricing can efficiently and effectively affect demand and thus impact GHG emissions additionally. This intermezzo discussion has explored aspects in which a long-term structural shift may differ from some of the aviation industry-based scenarios, which exclusively rely on the hope for technological progress. Achieving the kind of continuity that is needed to secure investment in building up the infrastructure for sustainable aviation and renewable energy systems poses some important challenges to political systems that tend to operate in short electoral cycles. While it is beyond the scope of this discussion—and will likely depend on national and regional circumstances to explain how such general challenges for long-term policies can be overcome—we have identified a number of specific ingredients for successful policies that would support a structural shift towards sustainable aviation. These include a balanced focus on technology and behaviour, a prominent role for price intervention on GHG emissions alongside the development of renewable energy sources, and policies that extend beyond R&D to support technology deployment as well as the creation of an enabling environment, which includes education and creating awareness among consumers. Additionally, the aviation industry could benefit from proactive self-regulation instead of waiting for policy actors to become active. To become really sustainable by 2050, the aviation industry needs to understand the risks of techno-optimism with respect to environmental problems and accept that technology alone will not be a free pass for endless growth and decreasing prices. Given the assumed growth rate of 300% in air transport demand until 2050 combined with the GHG reduction target of 50% in the same period, the feasibility of this GHG reduction target becomes highly questionable. Thus, it becomes clear that technology alone will not solve the emission problem of aviation by 2050. A successful and timely transformation towards full sustainability will require both, technological progress, and decreased growth in demand. The considerations of this chapter are important to especially understand how airlines should deal with this outlook and what it means for short- and long-term actions. This will be discussed in the upcoming chapters.

References

Alexander, S., & Rutherford, J. (2019). A critique of techno-optimism: Efficiency without sufficiency is lost. In *Routledge handbook of global sustainability governance* (pp. 231–241). Routledge.
ATAG. (2020). *Waypoint 2050 Report*. Retrieved April 07, 2021, from https://aviationbenefits.org/media/167187/w2050_full.pdf
Barry, J. (2016). *Bio-fuelling the hummer? Transdisciplinary perspectives on the transition to sustainability* (pp. 106–124). Routledge.
Dimitropoulos, J. (2007). Energy productivity improvements and the rebound effect: An overview of the state of knowledge. *Energy Policy, 35*(12), 6354–6363.
Gardezi, M., & Arbuckle, J. G. (2020). Techno-optimism and farmers' attitudes toward climate change adaptation. *Environment and Behavior, 52*(1), 82–105.
Hankey, S., & Marshall, J. D. (2010). Impacts of urban form on future US passenger-vehicle greenhouse gas emissions. *Energy Policy, 38*(9), 4880–4887.
Kirby, P., & O'Mahony, T. (2017). *The political economy of the low-carbon transition: Pathways beyond techno-optimism*. Springer.
Lee, D. S., Fahey, D. W., Skowron, A., Allen, M. R., Burkhardt, U., Chen, Q., Doherty, S. J., Freeman, S., Forster, P. M., Fuglestvedt, J., Gettelman, A., De León, R. R., Lim, L. L., Lund, M. T., Millar, R. J., Owen, B., Penner, J. E., Pitari, G., Prather, M. J., ... Wilcox, L. J. (2021). The contribution of global aviation to anthropogenic climate forcing for 2000 to 2018. *Atmospheric Environment, 244*, 117834.

Lerner, J. (1999). The government as venture capitalist: The long-run effects of the SBIR program. *Journal of Business, 72*, 285–318.

Rahm, D. (1993). US public policy and emerging technologies: The case of solar energy. *Energy Policy, 21*(4), 374–384.

Sanden, B. A., & Azar, C. (2005). Near-term technology policies for long-term climate targets - economy wide versus technology specific approaches. *Energy Policy, 33*(12), 1557–1576.

Spence, A., & Pidgeon, N. (2009). Psychology, climate change & sustainable bahaviour. *Environment: Science and Policy for Sustainable Development, 51*(6), 8–18.

Statista. (2021). *Durchschnittlicher Verbrauch von (versteuerten) Zigaretten pro Tag in Deutschland in den Jahren 1991 bis 2020.* Retrieved April 07, 2021, from https://de.statista.com/statistik/daten/studie/182391/umfrage/zigarettenkonsum-pro-tag-in-deutschland/

Teske, S., Pregger, T., Simon, S., Naegler, T., Graus, W., & Lins, C. (2011). Energy [R] evolution 2010—A sustainable world energy outlook. *Energy Efficiency, 4*(3), 409–433.

Tran, M. (2016). A general framework for analyzing techno-behavioural dynamics on networks. *Environmental Modelling & Software, 78*, 225–233.

Trutnevyte, E., Hirt, L. F., Bauer, N., Cherp, A., Hawkes, A., Edelenbosch, O. Y., Pedde, S., & van Vuuren, D. P. (2019). Societal transformations in models for energy and climate policy: The ambitious next step. *One Earth, 1*(4), 423–443.

Unruh, G. C. (2002). Escaping carbon lock-in. *Energy Policy, 30*(4), 317–325.

Von Weizsäcker, E., Lovins, A. B., & Lovins, L. H. (1998). *Factor four: Doubling wealth, halving resource use.* Earthscan.

Voß, J.-P., Smith, A., & Grin, J. (2009). Designing long-term policy: Rethinking transition management. *Policy Sciences, 42*, 275–302.

West, J., Bailey, I., & Winter, M. (2010). Renewable energy policy and public perceptions of renewable energy: A cultural theory approach. *Energy Policy, 38*(10), 5739–5748.

Williamson, K., Satre-Meloy, A., Velasco, K., & Green, K. (2018). *Climate change needs behavior change: Making the case for behavioral solutions to reduce global warming.* Rare.

Wüstenhagen, R. (2013, January 7). *Scenarios for a high penetration of renewable energy: Considerations about technology, behaviour change and policy.* Alexandria - University of St. Gallen. Retrieved from https://www.alexandria.unisg.ch/263408/

Introducing Sustainable Aviation Strategies

Judith L. Walls

Abstract

- Corporate strategy is about the capabilities companies develop to address shifts in the external business context, with the goal of outperforming peers in the marketplace.
- Corporate sustainability strategy embeds environmental and social goals into corporate strategy.
- To mitigate climate change, companies need to rethink both the business in which they operate, and the underlying business model.
- The aviation industry faces existential threat and reputation, regulatory, financial, and operational risks in the context of climate change.
- However, there are also many business opportunities in aviation, to transform businesses towards sustainability and gain competitive advantage through pioneering action.
- Corporate leaders are an essential part of driving sustainability transformation in aviation.

1 Corporate Sustainability Strategy in the Airline Industry

Strategy is the process by which companies set goals and priorities so that they can mobilize their resources with the goal of maximizing their performance. In a competitive marketplace, companies seek to outperform their peers by carving out their own space and creating value that customers are willing to recognize and pay

J. L. Walls (✉)
Institute for Economy and the Environment, University of St. Gallen, St. Gallen, Switzerland
e-mail: judith.walls@unisg.ch

© The Author(s), under exclusive license to Springer Nature Switzerland AG 2022
J. L. Walls, A. Wittmer (eds.), *Sustainable Aviation*, Management for Professionals,
https://doi.org/10.1007/978-3-030-90895-9_5

for. Not only do companies seek how to compete in markets, they also decide which markets to compete in.

Companies design their strategies based on two overarching factors: (i) the shifts and trends that are happening in the external business context and (ii) the resources, capabilities, and competences available inside the company. This helps companies to answer not only the "what" and "why" in shaping their strategy, but also the "how", "when", "where", and "who". In essence, companies seek to boost their own strengths—while managing their weaknesses—to mitigate threats and make the most of the opportunities that arise in the greater business context.

Shaping corporate strategy is not a one-off, static event. Rather, corporate strategy is a dynamic process that requires constant adjustment as the external business context experiences shifts and as the company's internal competences (e.g. innovation) develop. For many years, corporate strategy was divorced from social and environmental issues. However, this is no longer the case as social and environmental problems have become so widespread that business is not only being held responsible for their cause, but also for providing solutions. In the aviation industry—airplane manufacturers, airports, airlines—the overarching environmental issue is that of climate change.

The aviation industry is in the midst of experiencing a seismic shift to its business context both in the short- and medium- to long-term. In the short-term, the world is in the midst of a global pandemic that has led to an abrupt drop in flights above 95% in April 2020 and year-round (2020) 50% decrease in demand compared to 2019. In the medium- to long-term, the world is facing a climate crisis that needs to be urgently addressed. The aviation industry faces several risks when it comes to climate change: *regulatory risk* as policy changes are implemented; *financial risk* as investors and insurers lose their appetites for industries dependent on fossil fuels; *reputational risk* as customer preferences and perceptions change; and, *operational risk* as storms, flooding, and rising sea level affect ground and air operations. For example, by the end of 2020, some 1200 institutions (including pension funds, foundations, banks, and other organizations) divested USD 14 trillion in assets from the fossil fuel industry, or nearly 20% of the global value of all assets under management (Gofossilfree.org, 2021; Statistica, 2021). The fossil fuel industry, in other words, is rapidly losing its licence to operate. And as one of the direct carbon-emission generating industries, aviation may soon face similar questions of existential threat. Players in aviation industry will have little choice but to respond if they want to survive.

But whenever there are risks, there are also opportunities for businesses—those who adapt and transform can pioneer the way to a business-beyond-usual and benefit in the process. Such opportunities also arise from addressing climate change by taking a (circular) system or life-cycle perspective, along all stages of the aviation supply chain. Industry players who proactively anticipate the shift and transform their business strategies to position themselves strongly for the future. Such a transformation must be carefully planned, however, due to the long (technological innovation) lead times in the aviation industry.

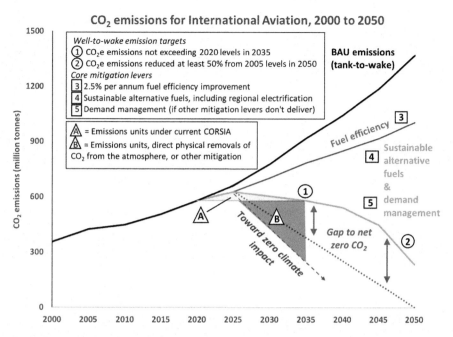

Fig. 1 CO₂ emissions for International Aviation, 2000–2050 (ICAO, 2019)

UN Secretary-General António Guterres called on leaders to reduce GHG emissions by 45% by 2030 and achieve net zero emissions by 2050 (UN, 2019). The Marrakech Partnership for Global Climate Action aims to implement these goals through multi-stakeholder collaboration. Specifically, many industries in the Marrakech Partnership identify climate action pathways by (i) reducing their use of raw materials and energy, (ii) increasing the productivity of energy and materials used, and (iii) decarbonizing production processes while transitioning to natural climate solutions (UN, 2020). In the aviation industry, ICAO (International Civil Aviation Organization) member states have agreed to achieving carbon neutral growth from 2020 and reducing emissions by 50% from 2005 baseline levels by 2050, with discussions on how best to reduce the gap to net zero CO_2 emissions and climate impact. The International Coalition for Sustainable Aviation (ICSA) has also produced enhanced climate mitigation targets and levers for international aviation for a pathway towards zero climate impact. Figure 1 shows ICSA's targets (1–2) and levers (3–5), and also predicts emissions units under the current CORSIA Scheme (A) if all international aviation emissions were covered, which they are not. Levers like fuel efficiency, alternative fuels, and demand management go a long way towards reducing emissions. However, additional mitigation actions (B) are needed to achieve zero carbon emissions in the aviation industry.

1.1 A Short Overview of Corporate Strategic Management

Companies can create value from their strategies in creating, manufacturing, or trading products and services. Value is then captured through creating (new) products and services, creating (new) markets, creating (new) methods of production and sourcing, or reorganizing the entire industry in a (new) way. From a strategy perspective, this means that companies have two levels at which they can achieve value capture (Porter, 1985) as seen in Table 1:

- *Corporate-level strategies*, for example, through diversification, horizontal and vertical integration, mergers and acquisitions, divestment, spin-offs, alliances and partnerships, and so on
- *Business-level strategies*, including low-cost, differentiation, niche, and hybrid approaches

2 Corporate Sustainability Strategy

When it comes to addressing sustainability and climate change, corporation embeds sustainability into their business strategy, throughout their operations including their supply chains and partnerships. Sustainability topics like climate change affect almost every part of a company's business: from raw material sourcing to production and manufacturing, selection of energy use, distribution, operational activities and

Table 1 Strategic options available to airlines at the corporate and business level

Type of strategy	Strategic options available to airlines	Examples from practice
Corporate level	Diversification	Passenger aircraft, cargo, logistics management, private charters
	Horizontal integration (including mergers and acquisitions)	Expanding range (short haul, long haul); acquiring other airlines
	Vertical integration (forward and backward)	Ticket booking (IT systems); ground handling; catering; car/hotel services
	Divestment/spin-offs	Lufthansa innovation hub spin-off start-up RYDES
	Alliances	Alliances such as oneworld, Star Alliance, SkyTeam; code-sharing, frequent flyer partnerships,
Business level	Differentiation	Full-service network airlines
	Low-cost	No frills point-to-point airlines
	Niche	Luxury, private charters, wetlease, and regional airlines
	Hybrid	Integration of point-to-point network with service differentiation, integration of global networks with regionally lower service levels

emissions from those activities, as well as consumer preferences, regulatory and stakeholder pressure. In the context of the aviation sector, these could, for example, include:

- Raw material sourcing: extraction of metals for airplane construction; agricultural practices of food sourced for onboard catering
- Production and manufacturing: energy use in constructing airplanes; food waste in putting together meals
- Energy use: fossil fuel versus renewable sources
- Distribution: transport of parts and components; relocation of crew to serve on flights; flight routes to serve customers
- Operational activities: fuel emissions of flights; food and plastic waste from serving passengers
- Consumer behaviour: emissions related to flying versus other modes of transport like rail; frequent flyers.

To integrate sustainability into their strategies, companies have moved away from linear economy take-make-waste models by which they take what they need from the natural environment and emit waste without considering the negative externalities. Instead, corporate strategy is increasingly shifting to a circular model in which the use of both biological and technical materials is designed in such a way as to eliminate waste and pollution and regenerate natural systems (Fig. 2). Specifically, biological raw materials come from renewable sources, at a rate no faster than the Earth can provide. Technical (man-made) materials are kept within the system through strategies like sharing, maintaining and prolonging, re-using and re-distributing, refurbishing and remanufacturing, or recycling products. The goal is to minimize leakage of waste that creates negative externalities.

In adopting a circular approach, corporations can create value through strategies like pollution prevention, product stewardship, clean technology, and collaboration across the supply chain including local communities (Hart & Milstein, 2003). In doing so, companies can gain a competitive advantage, for example, through becoming a first mover by pre-empting regulatory shifts and peers, maintaining legitimacy by integrating stakeholder concerns, gaining a strong future position, lower costs through improved efficiency, and developing an innovation competence by embedding sustainability solutions (Hart, 1995; Hart & Dowell, 2011).

The circular economy approach focuses both on energy and raw material (re)use and helps to address climate change, but companies also use more specific strategies for climate change which are discussed next.

3 Corporate Climate Change Strategy

Companies develop different types of strategies to help limit climate change and reduce the impact of climate change. Specifically, there are two types of strategies companies implement (IPCC, 2014): (i) *climate change mitigation* strategies that

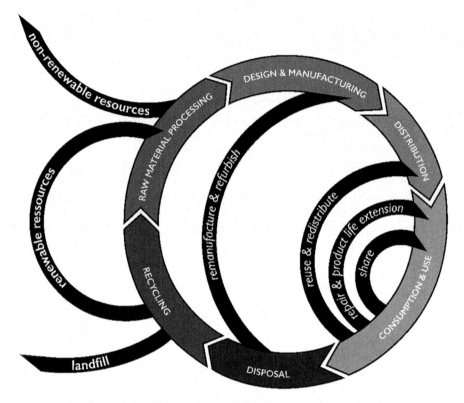

Fig. 2 The circular economy "value circle". © Zsusza Borsa (2021)

seek to reduce the underlying cause of climate change by reducing, limiting, preventing, and removing greenhouse gases; and (ii) *climate change adaptation* strategies to manage their vulnerabilities of the impacts of climate change by moderating or avoiding harm from climate change.

These strategies have separate purposes but are complementary as many firms pursue a combination of both strategies. However, they differ quite substantial in purpose and approach, as well as time horizon and spatial scale objectives. Mitigation strategies take a long-term perspective and a more global and international scale, whereas adaptation strategies tend to be shorter-term in outlook with a more local application (Table 2).

Climate *mitigation* strategies aim to reduce greenhouse gases drastically, focusing on the role of companies in addressing the climate problem. Sometimes climate mitigation is considered to be the task of governments rather than that of companies (Beermann, 2011). However, companies can benefit directly from such strategies as these reduce the reputational, financial, operational, and regulatory risks

Table 2 Types of corporate climate change strategies

	Mitigation strategy	Adaptation strategy	
Approach	Actions taken to reduce emissions that cause climate change	Adaptations made to operations to manage the impact of climate change	
Focus	Addressing underlying cause of climate change	Managing the risks and vulnerabilities due to exposure to climate change	
Sub-strategies	–	**Soft adaptation:** Refining existing processes and operations; incremental approach	**Hard adaptation:** Investing in new technology, infrastructure, and (re)location; transformational approach
Time horizon	Long-term	Short-term	Medium-term
Scale	Global	More local	
Financial investment	Medium-high	Low-medium	Medium-high
Examples in the aviation industry	Galápagos Ecological Airport designed to run completely on renewable energies (65% from wind, 35% from solar), and using 80% recycled materials in its construction and re-use of wastewater in its system (estimated costs USD 40 million (Egere-Cooper, 2015))	Changi Airport resurfacing Runway 1 with stronger and more durable asphalt with higher temperature resistance due to climate change (estimated cost USD50 million (Changi Airport, 2020))	Plans to build a new 10-mile long sea wall at San Francisco International Airport to guard against 3 feet of sea level rise plus 2 more feet for storms/waves (estimated cost USD587 million (Rogers, 2019))

(Averchenkova et al., 2016). As such, companies that are proactive preserve their legitimacy or licence to operate.

Climate *adaptation* strategy focuses more on how the company's own vulnerability (risks) to climate change can be managed. Corporate climate adaptation strategies range from incremental adaptation in which corporate activities do not change fundamentally (the so-called "soft" adaptation), to approaches with more fundamental changes ("hard" adaptation). A soft adaptation strategy is less involved, such as refining existing process or operations, and therefore more flexible. A hard adaptation strategy requires more permanent investments in new technologies, infrastructure, or relocation of facilities. An example of a hard adaptation strategy includes building on raised land to avoid flooding due to sea level rise. The benefit of engaging in a climate change adaptation strategy is that companies can seize business opportunities as demand preferences change, or improve efficiency and therefore profitability, while minimizing disruption to their production and processes

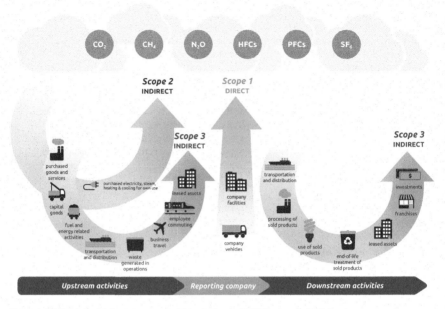

Fig. 3 GHG Protocol Scope 1, 2, and 3 emissions. GHG Protocol (2021)

(Averchenkova et al., 2016). However, adaptation strategies do not combat the underlying problems of industrial activity that drive climate change, and adaptation strategies can potentially accelerate climate change rather than slow it down (Lobell et al., 2013).

Without doubt, the most effective way to combat climate change is through mitigation strategies, which address the underlying cause of climate change. To do so, companies must both avoid and reduce emitting GHGs. Because companies do not operate in isolation but rather within a larger supply chain network, companies need to consider how to avoid and reduce their GHG emissions both directly and indirectly. The Greenhouse Gas Protocol (Fig. 3) is a framework that helps companies to identify the different areas along a company's operations and supply chain network where GHGs might be emitted, known as Scope 1, 2, and 3 emissions (GHG Protocol, 2021).

- Scope 1: Emissions created directly by the company as a result of its own operations and manufacturing, including facilities, and vehicles and other modes of transport, and so on.
- Scope 2: Indirect emission from purchasing electricity or steam, for heating and cooling, for company's own use.
- Scope 3: Indirect emissions generated along supply chain activities both upstream and downstream.

Introducing Sustainable Aviation Strategies

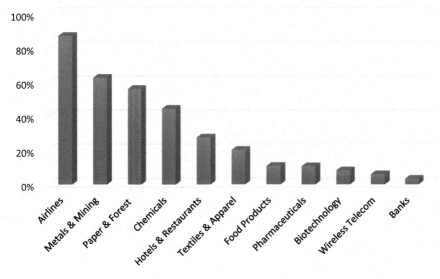

Fig. 4 Proportion of GHG emissions from Scope 1 sources—selected industries. Own illustration based on 2018 data from Trucost

- Upstream activities such as purchased goods and services, transportation, business travel and employee commuting, leased assets, fuel and energy related activities, and so on.
- Downstream activities such as transportation and distribution, any processing of sold goods down the line, use of sold goods, end-of-life (waste), investments and franchises, and so on.

Using data from Trucost, scope 3 emissions account for most of the GHG emissions across different industries. However, in the airline industry, scope 1 emissions represent the vast majority of GHG emissions at a whopping 87.3% (Fig. 4), compared to an average (mean) proportion of scope 1 emissions of 23.9% among all other industries, as a result of burning fossil fuels associated with flying. When considering only Scope 1 GHG emissions, airlines are the eighth highest emitter of these direct GHG emissions with 473 million tonnes of CO_2-eq emissions during 2018 (Fig. 5) which accounts for 3.9% of all scope 1 emissions across industries captured by Trucost. This figure is somewhat higher than those cited by the aviation industry, which may be due to differences in data collection. However, Lee et al. (2021) show that 2018 values of the aviation industry's contribution to climate change is 48% larger than their earlier study that uses 2005 data. As such, it is clear that as airline traffic continues to grow, the industry's contribution to climate change is also rising. Taken together, this means that not only is the airline industry a significant contributor to GHG emissions through its direct operations, but it also has a great deal of control over how to manage its emissions compared to other industries.

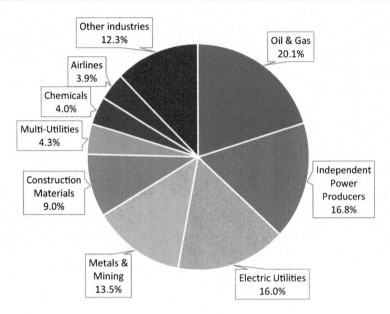

Fig. 5 Industry's proportion of all Scope 1 (direct) GHG emissions. Own illustration based on 5084 companies' data from Trucost (2018)

3.1 Climate Mitigation Strategies

Climate change mitigation strategies rely on actions companies (and others) take to reduce emissions that cause climate change. Similar to mitigation strategies that address the impact of industries on biodiversity loss (Mitchell, 1997), climate change (or "carbon") mitigation strategies are also hierarchical in their ability to effect meaningful change. In other words, simply mitigating climate change through offsetting actions is not going to make a meaningful difference in terms of the GHG emissions generated by the aviation industry. Instead, transformation of the industry as a whole in a manner that avoids emitting emissions in the first place should be the goal.

Avoiding emissions is, therefore, the first and most important step in any mitigation action as represented in the hierarchy of climate mitigation strategies in Fig. 6. What cannot be avoided is addressed in step two which is to reduce emissions, either by flying less or mobility through alternative modes of transport like (electric) train travel. The third step would be to replace traditional fossil fuels with cleaner alternatives that have fewer emissions such as biofuels or developing hybrid electric engines. The final step is offsetting for any remaining emissions. In some models, carbon capture is either an additional step after Offset, or a step in between Replace and Offset. We do not consider carbon capture here as a viable step because this technology is expensive (and well outside of the core capabilities of the aviation industry), is difficult to scale, and requires a centralized infrastructure approach (Wood, 2017).

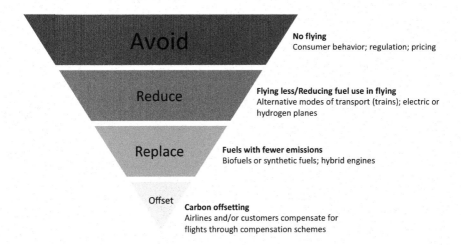

Fig. 6 Hierarchy of Climate Mitigation Strategies. The most effective strategy is to avoid emissions. Own Illustration

Each of these options clearly has implications for both corporate-level and business-level strategies of aviation companies, as well as the notion of circular economy in which the aviation industry must collaborate with other industries both upstream and downstream to achieve its climate mitigation goals.

3.2 Implications of Climate Change Mitigation Strategies for the Aviation Industry

The implications of climate change mitigation are substantial for corporate strategies in the aviation industry. For example, to avoid GHG emissions implies changes in consumer behaviour, regulation, or global pricing strategies with the implication that the overall demand for flying reduces (perhaps drastically). In this case, the industry would need to consider how to best manage such a decline in demand, or even exiting the industry in which it has invested an enormous amount of time and money. Such a scenario may be difficult for many stakeholders to accept psychologically. On the other hand, relying solely on carbon offsetting (the last step in mitigation) would be prohibitively costly. For example, with 473 million tonnes of CO_2-eq emissions during 2018, at a conservative carbon price of USD 50 per tonne (Fan et al., 2021), the total offsetting cost in 2018 was worth USD 23.7 billion.

A key question that players in the aviation industry should ask themselves is "what business are we in"? Typical answers may be: the "airline" or "airport" business. But reframing this answer by considering why people fly in the first place might be enlightening. People fly to connect with others for business or social purposes, to visit new places and broaden their horizons, for personal and spiritual

Table 3 Hierarchy of climate mitigation strategies including examples

Hierarchy of climate mitigation strategies	Examples of strategic options
1. Avoid emissions	Eliminate short-haul flight routes; leverage "service" expertise for trains/electric bus routes (see "reduce emissions") to meet travel demand on shorter routes Connect people in other ways (e.g. online teleconferencing for business customers) Forward integration with consumers for decision-making and nudging behaviour Co-develop regulation on avoiding flights/carbon credit schemes? Set carbon trading cap? (to level competitive playing field) Divestment from the airline industry Horizontal integration into electric shipping for cargo
2. Reduce emissions	R&D into electric, hydrogen and solar planes and hybrid engines Route planning for efficiency Optimizing loads, speed, altitudes; take-off and landing optimization Leverage alliances to plan fewer flights per week for specific routes Taxiing with electric vehicles (collaboration, backward integration) Horizontal integration into electric railway/bus lines
3. Replace emissions	Sustainable aviation fuels (biofuels and synthetic fuels), which are produced without fossil fuels, but still emit GHG upon use
4. Offset emissions	Develop IT platforms for offsetting with customers Full service; divest low-cost carriers to embed offsetting into ticket prices and service Enforce mandatory offsetting in the industry

growth, building and strengthening relationships, to escape, have adventures, relax, celebrate, and so on.

An example of how reframing the business industry players are in works can be taken from the automotive industry. Whereas in the past, the automotive industry was mainly about "selling cars", players in this industry reframed this perspective in anticipation of changes in mobility to autonomous, connected, electric, and shared (ACES) vehicles. In acknowledging this shift, the automotive industry is no longer about car ownership but rather about transportation and mobility. This insight can help to inform the strategic directions aviation industry players should consider if they want to remain relevant. Would it be unfeasible, for example, to consider that an airline engages in a horizontal integration strategy to purchase an electric railway business? What about diversifying into online conferencing platforms? Or building spas and adventure parks, or couples retreats? Examples of how such questions might be relevant to the aviation industry are provided in Table 3.

To pursue these strategies, aviation companies would likely need to build new capabilities and competences for climate change and sustainability. In embedding such skills into the company's processes and operations, the benefit is resulting flexibility and innovation (Busch, 2011) that allows companies to think in new ways and bring new approaches to the industry (Winn et al., 2011). Engaging in risk assessment and scenario planning is one way by which corporate leaders and other members can create awareness of climate change and develop strategies (Haigh, 2019).

4 Sustainability Strategies for Managing Consumers (and Consumer Demand)

While managing sustainability upstream requires substantial time, effort, and coordinated strategic action for companies, it is typically also easier to do than managing the last step in the product life cycle which is consumer behaviour. Yet, closing the production or supply chain loop requires industries to involve consumers in the process. A complete circular economy begins with the design of a product (or service) life cycle and builds in loops to intervene in the product use phase to avoid waste (see Fig. 2 on circular economy). The key concept here is to focus on *creating value* for the consumer, and *prolonging* that value created, in a manner that minimizes production and consumption. In a circular economy system, such value can be created in two ways (Bocken et al., 2016):

- By closing the loop of resource flows: designing for material (technical) and biological recovery, re-use in remanufacturing, and similar
- By slowing the flow of resources in the loop: designing products that last longer, can be serviced/fixed/repaired, and encouraging a sufficiency approach

In other words, instead of seeking to grow the entire market (for the product or service), in a sustainable strategy the emphasis is on growing the *value* of the product or service. In doing so, customers may switch from products or services that create less sustainable value, thus growing a business by capturing market share from others. And, where possible, the market for the product or service may even experience degrowth (but not create less value or profit) as the frequency of demand or overall demand declines because the product or service lasts longer or provides a different kind of (more sustainable) benefit.

Companies have different options when seeking to create value. For example, for airlines, the following sources of growth exist:

(a) First, airlines can *grow their overall market*. Pre-pandemic, ICAO estimated an average 4.3% increase per year in air transport demand for the next 20 years, with the majority of new demand coming from Asia. However, such a growth implies rising greenhouse gas emissions from the airline industry, and therefore the industry would not meet the Paris Agreement targets.
(b) Second, airlines could provide value to consumers in switching to *alternative, more sustainable* transport options. This could, for example, be in the form of business-level strategies such as differentiation, by offering higher priced flights but using more sustainable forms of fuel or aircraft. Over time, traditional

fuel-based aircrafts might be entirely replaced with renewable energy aircraft. Airlines might achieve such an option via a corporate-level strategy of partnering with technology start-ups in this space.

(c) Third, airlines could consider *shrinking the overall market* for flights, a so-called degrowth strategy. This might take the form of a corporate-level strategy in which airlines partner with ultra-high-speed rail companies for regional routes and offering rail travel as an alternative to flights. Consumers, essentially, still gain the same value from the product (mobility from point to point), but via an alternative option than flying.

In options (b) and (c), as consumer preferences for sustainable alternatives rise (perhaps in conjunction with awareness campaigns and education), the airline industry might experience consolidation and shake-out. Less sustainable airlines might be acquired and converted towards more sustainability, for example. Or less sustainable airlines might exit the market altogether if they can no longer make a profit.

5 Leading the Change Towards Sustainable Aviation: The Role of (Top) Managers

Leaders are an essential driver of creating organizational change for sustainability. CEOs are well aware of the need to address topics like climate change, identifying not only that sustainability is critical to business success, but also that businesses are not doing enough (Gupta, 2019). Scholarly research is also increasingly acknowledging the importance of leaders such as CEOs in corporate sustainability. For example, CEOs are responsible for nearly a third (28%) of a company's environmental performance, with much smaller proportions of attributed to other dynamics such as industry (7%), company-specific characteristics (21%), or time (10%) (Wernicke et al., 2021). Clearly, the role of leaders in corporate sustainability is highly relevant.

5.1 What Drives Leaders to Engage in Sustainability?

Leaders can play a role in corporate sustainability for several reasons (Salaiz et al., 2020). First, leaders' personal motivations matter, rooted in: (i) *instrumental drivers*, based on self-interest such as, for example, seeking a financial benefit to the company or benefitting personally via compensation reward systems, (ii) *relational drivers*, based on the desire to maintain good relationships with stakeholders and show responsible action towards society, and (iii) *moral drivers*, based on values orientation that sustainability is the "right thing to do".

Second, leaders' characteristics and traits matter. Leaders that engage in sustainability may have stronger sense of duty to take action on climate change, and also be able to handle more complexity in decision-making. Personal experience with sustainability topics also makes a big difference as this helps to focus attention on both the problems and opportunities that sustainability challenges create. These types of traits or characteristics act as a filter through which leaders interpret the world around them and may lead them to pay more attention to sustainability.

Other characteristics likely also matter (Walls et al., 2020). For example, emotions such the love or fear of nature may explain why some leaders have a stronger, positive attachment to nature than others. Values such as concerns to protecting the natural environment are similarly important. While values differ across individuals and culture, four values—health, well-being, longevity, and environmental protection—are universal across all cultures and these can inform how corporate leaders can create change within their companies that internal and external stakeholders buy into (Walls & Triandis, 2014).

5.2 What Types Leaders Drive Positive Change for Sustainability?

Trailblazing a new path for sustainability in a company is not an easy task. One question that arises is how can companies identify, develop, and empower their leaders to take action on sustainability? Insights from research show that three distinct levels of characteristics are relevant (Walls et al., 2020):

1. *Individual level characteristics*: leaders who are optimistic and humble have compassion, are competent, and reliable. Such leaders have the courage to create change and are also resilient to overcome obstacles.
2. *Relational level characteristics*: leaders who are conscientious and authentic. Such leaders are able to build strong networks and become strong negotiators.
3. *Collective level characteristics*: leaders who embrace paradox and have the ability to reframe challenges or conflict as opportunities. Such leaders can leverage small wins and organize collective action (e.g. internally, among peers/industry, etc.).

Leaders who have these characteristics are those who successfully drive positive change. That is not to say that a leader must be born with them. Rather, many of these characteristics can be developed to help empower leaders to engage in sustainability action.

In addition, the more influence the leader has over others in the organization, the better, both in the form of formal control over the management team, but also informal influence such as through a leader's personality or experience (Walls & Berrone, 2017).

5.3 How Can Leaders Frame Sustainability to Encourage the Entire Company to Change?

Leaders can *frame* sustainability in different ways to both internal and external stakeholders. One framing approach is a *business case* for sustainability, for example, by arguing that it makes financial sense as well as benefiting the natural environment. Such an approach may have strong benefits, especially in a company or sector that has traditionally been resistant to engaging in sustainability strategies.

A second option is to make a *moral case*, by arguing that sustainability is simply the right thing to do, for example, by arguing that it is an issue of saving the planet or one of environmental justice. This type of framing may work in some settings, but also can meet with a lot of resistance when values differ between stakeholders.

A third option is to make a *holistic case*, by arguing that everything is connected, for example, that business depends on the natural environment as much as it benefits from it. The benefit of a holistic case is that it takes adopts an "and-both" mindset rather than an "either-or" mindset. However, a drawback is that a holistic approach requires more comfort with uncertainty and complexity, which may be more difficult to do for individuals with less sustainability experience.

No one framing approach is superior to the others. But some approaches may resonate better with internal and external stakeholders than others, depending on the company's unique situation and context. Important is that the leaders embed their sustainability strategy throughout the company's structures, systems, and processes. For example, leaders who create a strong vision for sustainability, and set in place supporting incentive structures, oversight and monitoring, reporting lines to the board, management and operational systems, and so on, will be much more likely to succeed (Walls et al., 2011). Similarly, it is important to get external stakeholders such as shareholders on board, by attracting investors with a long-term outlook who are more likely to be supportive of long-term climate change strategies (Walls et al., 2012).

6 Conclusion

In the context of climate change, players in the aviation industry face an existential threat. Companies must transform their business, due to the reputational, regulatory, financial, and operational risks that climate change brings. A key part of addressing climate change is to embed sustainability into the core part of aviation players' business, through a circular business model approach and mitigating action on climate change in which offsetting should be the last step. The first step is to ask: "what business are we in?" In answering this question, aviation companies can identify opportunities for sustainability transformation to ensure future, long-term competitiveness. This chapter identifies both corporate- and business-level sustainability strategies available to aviation companies. Corporate leaders play an essential role in driving this strategic transformation and can pave a pioneering path towards sustainability.

References

Averchenkova, A., Crick, F., Kocornik-Mina, A., Hayley, L., & Surminski, S. (2016). Multinational and large national corporations and climate adaptation: Are we asking the right questions? A review of current knowledge and a new research perspective. *WIREs Climate Change, 7*(4), 517–536.

Beermann, M. (2011). Linking corporate climate adaptation strategies with resilience thinking. *Journal of Cleaner Production, 19*(8), 836–842.

Bocken, N. M. P., de Pauw, I., Bakker, C., & van der Grinten, B. (2016). Product design and business model strategies for a circular economy. *Journal of Industrial and Production Engineering, 33*(5), 308–320.

Busch, T. (2011). Organizational adaptation to disruptions in the natural environment: The case of climate change. *Scandinavian Journal of Management, 27*(4), 389–404.

Changi Airport. (2020). *Enhancing safety and productivity at Changi Airport's runway*. Retrieved April 12, 2021, from https://www.changiairport.com/corporate/media-centre/changijourneys/the-airport-never-sleeps/enhancing-safety-and-productivity-at-changi-airports-runway.html

Egere-Cooper, M. (2015, August 4). This airport runs on wind. *Business Traveller*. Retrieved April 12, 2021, from https://edition.cnn.com/travel/article/galapagos-ecological-airport-wind-and-solar-power/index.html

Fan, J., Rehm, W. & Siccardo, G. 2021. *The state of internal carbon pricing*. McKinsey & Company. Retrieved April 12, 2021, from https://www.mckinsey.com/business-functions/strategy-and-corporate-finance/our-insights/the-state-of-internal-carbon-pricing

GHG Protocol. (2021). *You, too, can master value chain emissions*. Greenhouse gas Protocol. Retrieved April 11, 2021, from https://ghgprotocol.org/blog/you-too-can-master-value-chain-emissions.

Gofossilfree. (2021). *Divestment Commitments*. Retrieved April 11, 2021, from https://gofossilfree.org/divestment/commitments/

Gupta, A. (2019). The decade to deliver: A call to business action. The United Nations Global Compact—Accenture Strategy CEO Study on Sustainability 2019.

Haigh, N. (2019). *Scenario planning for climate change: A guide for strategists*. Routledge.

Hart, S. L. (1995). A natural-resource-based view of the firm. *Academy of Management Review, 20*, 986–1014.

Hart, S. L., & Dowell, G. (2011). A natural-resource-based view of the firm: Fifteen years later. *Journal of Management, 37*(5), 1464–1479.

Hart, S. L., & Milstein, M. B. (2003). Creating sustainable value. *Academy of Management Perspectives, 17*(2), 56–67.

International Civil Aviation Organisation [ICAO]. (2019). *Envisioning a « zero climate impact" international aviation Pathway towards 2050: How governments and the aviation industry can step-up amidst the climate emergency for a sustainable aviation future*. Working Paper.

IPCC. (2014). Annex II: Glossary. In K. J. Mach, S. Planton, & C. von Stechow (Eds.), *Climate change 2014: Synthesis report. Contribution of working groups I, II and III to the fifth assessment report of the intergovernmental panel on climate change*. Geneva, Switzerland: IPCC, pp. 117–130.

Lee, D. S., Fahey, D. W., Skowron, A., Allen, M. R., Burkhardt, U., Chen, Q., Doherty, S. J., Freeman, S., Forster, P. M., Fuglestvedt, J., Gettelman, A., De León, R. R., Lim, L. L., Lund, M. T., Millar, R. J., Owen, B., Penner, J. E., Pitari, G., Prather, M. J., ... Wilcox, L. J. (2021). The contribution of global aviation to anthropogenic climate forcing for 2000 to 2018. *Atmospheric Environment, 244*, 117834. https://doi.org/10.1016/j.atmosenv.2020.117834

Lobell, D. B., Baldos, U. L. C., & Hertel, T. W. (2013). Climate adaptation as mitigation: The case of agricultural investments. *Environmental Research Letters, 8*, 1.

Mitchell, J. (1997). Mitigation in environmental assessment – Furthering best practice. *Environmental Assessment, 5*(4), 28–29.

Porter, M. E. (1985). *Competitive advantage: Creating and sustaining superior performance.* Simon & Schuster.

Rogers, P. (2019). *SFO plans to surround airport with 10-mile wall to protect against rising bay waters.* Retrieved April 12, 2021, from https://www.mercurynews.com/2019/10/10/sfo-plans-to-surround-airport-with-10-mile-wall-to-protect-against-rising-bay-waters/

Salaiz, A., Chiu, S. C., & Walls, J. L. (2020). Sustainability agency at the top of the organization: Microfoundations research on corporate sustainability. In S. Teerikangas, T. Onkila, K. Koistinine, & M. Kel (Eds.), *Edgar Elgar research handbook of sustainability agency.* Edward Elgar.

Statistica. (2021). *Global assets under management.* Retrieved April 11, 2021, from https://www.statista.com/statistics/323928/global-assets-under-management/

UN. (2019). *2019 climate action summit.* United Nations Climate Action. Retrieved April 11, 2021, from https://www.un.org/en/climatechange/2019-climate-action-summit

UN. (2020). *Climate action pathway – Industry. Executive Summary.* United Nations Climate Change, Global Climate Action & Marrakesh Partnership. Retrieved April 11, 2021, from https://unfccc.int/sites/default/files/resource/ExecSumm_Industry_0.pdf

Walls, J. L., & Berrone, P. (2017). The power of one to make a difference: How informal and formal CEO power affect environmental sustainability. *Journal of Business Ethics, 145*(2), 293–308.

Walls, J. L., Berrone, P., & Phan, P. H. (2012). Corporate governance and environmental performance: Is there really a link? *Strategic Management Journal, 33*(8), 885–913.

Walls, J. L., Phan, P. H., & Berrone, P. (2011). Measuring environmental strategy: Construct development, reliability, and validity. *Business & Society, 50*(1), 71–115.

Walls, J. L., Salaiz, A., & Chiu, S. C. (2020). Wanted: Heroic leaders to drive the transition to 'business beyond usual'. *Strategic Organization.* https://doi.org/10.1177/1476127020973379

Walls, J. L., & Triandis, H. C. (2014). Universal truths: Can universally held cultural values inform the modern corporation? *Cross Cultural Management. An International Journal, 21*(3), 345–356.

Wernicke, G., Sajko, M., & Boone, C. (2021). How much influence do CEOs have on company actions and outcomes? The example of corporate social responsibility. *Academy of Management Discoveries.* https://doi.org/10.5465/amd.2019.0074

Winn, M. I., Kirchgeorg, M., Griffiths, A., Linnenluecke, M., & Günther, E. (2011). Impacts from climate change on organizations: A conceptual foundation. *Business Strategy and the Environment, 20,* 157–173.

Wood, T. (2017). *The challenges for CCS.* Grattan Institute. Retrieved June 4, 2021, from https://grattan.edu.au/news/the-challenges-for-ccs/

Airline Perspective

Juliette Kettler and Judith L. Walls

Abstract

- Current proposals to mitigate emissions from air travel by IATA will be insufficient to meet Paris Agreement Goals. More drastic transformation in the airline industry is needed.
- Airlines can benefit by engaging in climate mitigation strategies, which represent strategic opportunities both in the short- and long term.
- The industry has made progress on emission reductions by efficiency gains in technology, route optimization, and aircraft design. However, future gains in these areas are likely to be marginal only.
- The most promising strategies are in "consistency" and "sufficiency." The consistency strategies focus on replacing kerosene with synthetic fuels in the long term (and biofuels in the short term). The sufficiency strategy focuses on reducing (demand for) flights, for example, by partnering with high-speed rail, especially on short-haul routes and raising ticket prices.
- Policy support is needed for both consistency and sufficiency strategies to scale up and commercialize synthetic fuels and other alternative technologies and level the competitive playing field for ticket prices.
- The upside-down pyramid model can help airlines to identify the most impactful climate mitigation strategies, starting with "avoid," then "reduce," "switch," and finally "offset."
- Offsetting through compensation is a last resort option available to airlines, as offsetting schemes are typically fraught with problems and do little to mitigate

J. Kettler (✉)
University of St. Gallen, St. Gallen, Switzerland

J. L. Walls
Institute for Economy and the Environment, University of St. Gallen, St. Gallen, Switzerland
e-mail: judith.walls@unisg.ch

climate change. A key factor for success is selecting the right compensation schemes.

1 The Business Case for Sustainability in Airlines

Airlines play a key role when it comes to environmental sustainability in the aviation industry. Not only do they emit massive Scope 1 greenhouse gas emissions as part of their day-to-day operation, they also face increasing pressure from stakeholders on climate change action. Business leaders in the airline industry can view climate change action as a risk mitigation exercise or as business opportunity.

From a *risk perspective*, managing climate change is about avoiding the loss of legitimacy, or image and reputation of a company and the license to continue to operate. A risk perspective tends to focus on demands from stakeholders like investors, policy makers, customers, the media, and activists. For example, the #Fridays4Future movement's "flygskam" (flight shame) message has created stigma around flying and led to significant decline in airline passengers and increase in rail travelers across Europe (Chap. "Perceptions of Flight Shame and Consumer Segments in Switzerland") and climate change is considered to be a top priority for airlines according to 70% of UK passengers (NATS, 2020). By actively managing sustainability, companies lower their unsystematic risk and reduce other negative consequences such as costs associated with emissions fines and violations and loss of reputation and legitimacy (Bansal & Clelland, 2004).

From an *opportunity perspective*, managing climate change creates value for the company, as companies build internal capabilities (e.g., technology, design, production, distribution, procurement, supply chain management) that generate a competitive advantage through lowering costs, preempting competitors and establishing a future position in the industry (Eccles et al., 2014; Hart, 1995). Companies who take this perspective have stronger brands, more cost savings through operational efficiencies, improved customer loyalty, additional revenue sources and greater employee productivity (Berns et al., 2009, p. 15). The value created from such actions is also among investors who doubled their investments and saw a 30% rise in available sustainability funds from 2019 to 2020. Sustainability fund investments now represent one third of the total assets under management in the USA (Morningstar, 2021).

In light of this growing pressure from stakeholders and competitors, it becomes clear that having a long-term strategy that entails sustainability is no longer a "nice marketing add-on" but an essential, must-have element that is critical in securing its long-term success. Airlines need to recognize that the environment is changing, and if they do not change with it, they will become obsolete.

1.1 Creating Value from Sustainability

But how can airlines create value from sustainability? As business-to-consumer (B2C) companies and large contributors to climate change, airlines have enormous potential to create value from climate action. In fact, airlines that include sustainability initiatives in their marketing differentiate themselves from peers and gain a competitive advantage (Mayer et al., 2014). At the same time, it is important that airlines consider the role of uncertainty and risk because of the unique context in which the industry operates.

The external business context for airlines is often described as volatile, uncertain, complex, and ambiguous (abbreviated as VUCA). Airlines have faced many external shocks that greatly impacted the industry, such as 9/11 in 2001, the financial crisis of 2008, and major shifts in the industry, such as the deregulation in the airline market in the USA (Wittmer et al., 2021). More recently, COVID-19 has disrupted not only airlines but industries globally (Linden, 2021). In the context of a VUCA setting, how can airlines move forward strategically and sustainably?

One way to gain long-term competitive advantage in spite of uncertain contextual settings is to build dynamic capabilities, defined as a company's "ability to integrate, build, and reconfigure internal and external competences to address rapidly changing [business] environments" (Teece et al., 1997). A central element of this approach is to introduce uncertainty as a standard factor, build *sensing*, *seizing* and *transforming* assets, and "get comfortable with the unknown" (Johnson, 2015). Management of uncertainty does not fixate on one clear-enough or a few alternate futures, but instead acknowledges ambiguous futures, i.e., a high number of possible futures. Common planning methods in this approach are scenario planning, backcasting, or pre-mortem analyses (Linden, 2021). For example, Finnair successfully applied scenario planning in dealing with the COVID-19 crisis, allowing the airline to quickly adapt to the new situation. As the COVID-19 situation unfolded, the airline based its decisions with reference to the earlier SARS epidemic, outlining three scenarios for industry recovery and engaging in financial planning based on the worst-case scenario (Trocmé, 2020).

To successfully navigate uncertainty, companies need to constantly scan external developments and challenge previously established assumptions. If significant changes or new opportunities arise, organizations can then be agile and adapt their structure and strategic assets accordingly (Teece et al., 1997, 2020). Being open and curious about change in the business environment and embracing as well as managing uncertainty proactively will help aviation managers to create more resilient organizations while simultaneously engaging in long-term planning.

Building dynamic capabilities proactively means that companies do not need to wait for regulatory bodies to design sustainability policies for more certainty in the business environment. Rather, airlines can act right away. In doing so, managers of airlines will develop agile organizations that can adapt to shifts and shocks as they arise. An added advantage is that airlines can thereby tackle climate change before it is too late.

Technological progress	Improved Infrastructure	Operational measures	Economic Measures
- Innovation in aircraft and engine technologies - Alternative fuels	- Improved use of airspace - Airport infrastructures adapted to needs	- More efficient aircraft sizes - Optimal flight routes and speeds - Optimized processes on the ground	- A global, sensibly designed, market-based system for reducing emissions to complement the other three pillars

Fig. 1 Overview of the Four-Pillar Strategy to mitigate climate change. Illustration adapted from IATA (2021)

1.2 Current Climate Change Approaches by Airlines

Airlines are no strangers to the ecological impacts of their businesses and the challenges that these bring. For example, the Lufthansa Group and British Airways have implemented sustainability and responsibility initiatives since the 1980s. Even so, it is clear that GHG emissions are becoming an increasingly urgent topic for airlines to address. In 2009, the International Air Transport Association (IATA), whose 299 member airlines carry 82% of the world's total air traffic (IATA, 2018), adopted three goals to mitigate CO_2 emissions:

- An average improvement in fuel efficiency of 1.5% per year from 2009 to 2020
- A cap on net aviation CO_2 emissions from 2020 (carbon neutral growth)
- A reduction in net aviation CO_2 emissions of 50% by 2050, relative to 2005 levels

To achieve these goals, the industry follows a so-called "basket of mitigation measures," encompassing four pillars with the purpose of combatting climate change (Fig. 1).

Technological progress measures aim at introducing greener aircraft technology. This is done by improving the efficiency of kerosene-powered aircraft as well as by introducing new engine technologies and alternative fuels. Efficiency measures have already contributed to lowering emissions. For example, between 2003 and 2018, SWISS managed to decrease its fuel consumption by 29% with efficiency measures (SWISS, n.d.), exceeding the annual fuel efficiency improvement target of 1.5%. Fuel efficiency is easier to achieve when airlines have young fleets. For example, Emirates increased their efficiency by 3–4% simply through folded wingtips, a design innovation of young aircraft models (Setchell, 2018).

While efficiency measures are being implemented, innovation in engine technology is still in a developmental/exploration phase. For example, the first solar-powered flight around the world, carrying only the two pilots, was completed in 2016 (Solar Impulse Foundation, 2016). Much research is being done on the possibility of electric, hybrid, and hydrogen engines, supported by aircraft manufacturers, tech companies, start-ups, and independent research groups. Such research and development represent investment opportunities for airlines. For example, easyJet collaborates with the start-up Wright Electric (easyJet, 2019). Airbus is similarly engaged in a hydrogen-powered zero-emissions aircraft that can carry up to 200 passengers, called ZEROe, set to enter the market by 2035 already (Airbus, 2020). Nevertheless, large-scale electric and hydrogen aircrafts remain a distant future goal (see Chap. "Technology Assessment for Sustainable Aviation").

Another new development is sustainable aviation fuels (SAF). There are two types of SAFs. Biofuels can be produced from a variety of biomass and feedstock like woodchips, coconuts, algae, domestic waste or other sources. Synthetic fuels, on the other hand, are produced from water, renewable energy (often wind or solar power) and CO_2 from the atmosphere. Both types of SAF are already used by airlines in common aircrafts. The advantage of biofuel is that it is commercially available, and many airlines have started to invest in this technology: the Lufthansa Group has made use of biofuels since 2011; Qatar Airways and Etihad Airways both launched biofuel research and development initiatives (Lufthansa Group, 2019; Setchell, 2018). However, since biofuels compete with food production, the industry has turned its attention to the relatively new technology of synthetic fuels.

Improved infrastructure encompasses the optimization and shortening of flight routes and procedures. The European airspace is especially inefficient due to the high number of air traffic control centers and different national systems. The Single European Sky (SES) initiative aims to harmonize the European airspace (see Chap. "The Role of Public Policy") in which various airlines participate. An innovative example is the Virtual Center Model by Skyguide, Switzerland's air navigation service provider, which unites processes, applications, and operations of air traffic service units in different locations. In another example, Emirates has developed partnerships to work towards shorter and more efficient flight paths across the globe (Emirates, 2021).

Other infrastructure improvement areas include optimizing landing approaches, airport infrastructure, and linking air travel with rail and road traffic. In Switzerland, there are specific Airtrains, for example from Lugano or Basel to Zurich Airport. The ticket for this train is included in the flight ticket with SWISS (2020).

Operational measures aim to continuously improve the fuel efficiency of daily operations in the air as well as on the ground through flight planning (e.g., concerning aircraft size, load factor altitude and speed), weight reduction (e.g., through lighter materials), or taxi-in processes after landing. These measures can extend beyond aircrafts: The Lufthansa Group aims at CO_2-neutral operations on the ground in all its home markets (Germany, Austria, and Switzerland) by switching to electric vehicles and using energy from renewable sources for buildings and vehicles (Lufthansa Group, 2019).

Operational measures also include reducing the use of other resources like paper, plastic, or food. The Lufthansa Group launched its "Flygreener" initiative in 2012, aimed at reducing waste on board, optimizing loading, and increasing recycling (Lufthansa Group, 2019).

Economic measures are about Global Market-Based Measures (GMBM) to help the industry reach its sustainability goals more quickly. In contrast to the other three pillars, this measure is aimed less at the long-term goal of reducing emissions and more at the short-term goal of compensating for emissions with the aim of achieving carbon neutral growth by 2020.

In 2016, ICAO Member States developed and agreed upon implementing a global Carbon Offsetting and Reduction Scheme for International Aviation, or CORSIA. CORSIA's goal is to offset all aviation sector emissions that cannot be reduced via other pillars by the 2020 target. A downside is that participation in CORSIA is voluntary until 2026. Nevertheless, 80 states intend to participate in CORSIA, representing a large share of the industry. By participating in CORSIA, airlines invest heavily in climate change action (ICAO, 2019). In addition, CORSIA allows for air travel ticket prices to be raised globally without distortion through national policies that creates comparative and competitive disadvantages. Higher ticket prices could result in decreased demand (and stronger profit margins), thereby reducing emissions. Finally, CORSIA provides an incentive for airlines to reduce emissions and develop new, sustainable technologies.

Airlines should, however, not overestimate the effectiveness of market-based measures. A study comparing the CO_2 emissions reduction potential of EU ETS and CORSIA (by 2030) against a scenario where no climate action is taken showed that the reduction potential for EU ETS is between -0.3% and -3.1% and for CORSIA between -0.4% and -0.7% (Peter et al., 2016). These numbers show how far market-based measures are from achieving the needed 100% reductions. Reasons for the lack of impact of market-based measures is that the low price of compensation schemes makes long-term investment into other solutions less attractive. In addition, CORSIA ignores the Radiative Forcing Index (RFI), so *only half* of the aviation industry's effect on climate change is compensated through CORSIA.

Next to CORSIA, airlines also take other measures to offset emissions. For example, easyJet compensates the CO_2 emissions for all its flights without involving passengers in the process (easyJet, 2020), and British Airways has offset all its flights within the UK since January 2020 (British Airways, 2020). Finnair and the Lufthansa Group integrate CO_2 compensation into the booking process (Finnair, 2020; Lufthansa Group, 2019), and offer not only offsets that invest in certified climate protection projects, but also into biofuels for the specific route in question, to be used in future flights. The Lufthansa Group also offsets all official staff air travel and shoulders 50% of compensation costs for some corporate customers.

Compensation schemes such as offsetting emissions have several limitations. First, only 1–2% of passengers compensate their flights. Second, many other industries also want to offset their emissions—but there is an insufficient number of projects to accommodate all sectors. Third, compensation is often done in developing countries which not only raises ethical questions but also questions of monitoring and oversight of the projects. Fourth, compensation schemes can create a rebound effect: people feel better about flying because they offset emissions and end up flying more. Finally, compensation projects like reforestation require decades before CO_2 emitted is fully compensated. For example, offsetting emissions, for a flight from Zurich to Singapore, via reforestation takes 20 years to be achieved (Compensaid, 2020).

These arguments all point to the conclusion that compensating emissions is not *the* solution to the decarbonization of aviation. Nevertheless, compensation can be *part* of the solution. Programs like CORSIA will likely result in a considerable share of emissions being compensated to help the airline industry to achieve climate neutrality targets. However, because of the limitations of compensation schemes, these should be considered as an interim solution and the last resort step in a series of actions to reduce emissions (see Chap. "Introducing Sustainable Aviation Strategies"). Bridging the gap to climate neutrality requires airline managers to focus on solutions that reduce actual emissions rather than compensating for them.

In conclusion, airlines are already engaging in approaches to reduce their impact on climate change. Many airlines aim to go beyond the IATA strategy. Delta Air Lines, for example, aims to be the first carbon neutral airline globally and mitigate all its emissions from March 2020 onwards (Delta, 2020) through efficiency measures, sustainable aviation fuels, and removing carbon from the atmosphere through forestry, wetland restoration, and other nature-based solutions.

However, current actions taken by the airline industry are insufficient at a global level. Overall emissions from airlines have rising by one third of volume growth, driven by an increase in traffic tripled over the past two decades (Boeing, 2019). This is a worrying development, as it counters the goals set by the Paris Agreement. As such, the strategy set by IATA is unlikely to limit global warming to below 2 °C. To meet climate mitigation targets, 100% reduction of fossil resource use and net neutrality concerning greenhouse gas emissions in the airline industry is necessary by 2050. However, the proposed strategy aims only to halve the emissions of 2005 by 2050. In addition, the IATA strategy focuses largely on efficiency and technological solutions and not at all on behavior change to reduce consumer demand for flights or shift consumers to more sustainable modes of transport.

2 Long-Term Strategic Sustainability in the Airline Industry

2.1 Insights from COVID-19 on Long-Term Aviation Development

Before COVID-19 spread through the world, air travel was predicted to grow for many decades to come, alongside a drastic increase in emissions. ICAO (2016) estimated that CO_2 emissions from aviation would double, possibly even triple, from 2020 to 2050. However, in light of the pandemic, new predictions are that passenger numbers will not return to pre-pandemic level until 2024 (IATA, 2020c). As a result, airlines have had to adjust their strategic goals. For example, Lufthansa has set new goals to reduce costs by 15%, reduce its fleet by 100 aircraft, and eliminate 22,000 full-time jobs by 2023 (Lufthansa Group, 2020b). As of September 2020, these goals were expanded to 150 aircraft and 27,000 jobs (Zeit, 2020).

The pandemic creates both constraints and opportunities for airlines. On the one hand, many airlines face liquidity bottlenecks and the need to implement austerity measures. On the other hand, this is an opportunity for airlines to take a fresh perspective on the industry in light of sustainability in mobility. For example, the French government rescue package that supports Air France mandates that the airline cannot operate flights for routes that take less than 2.5 h by train. Also attached to the package is the modernization of its fleet in order to reduce CO_2 emissions by 50% per passenger kilometer by 2030 (Aerotelegraph, 2020).

In light of the Paris Agreement and stakeholder pressure, climate change will remain a highly relevant topic for airlines in the coming years. Nevertheless, airlines still operate on the basis of driving passenger growth, particularly in the Asian market (Boeing, 2019). For example, flight subscription models in China costing as little as CHF500 per year encourages demands for flights that would not otherwise be needed, according to Professor David Yu, finance and aviation expert at NYU University in Shanghai (SRF, 2020). Even pessimistic forecasts predict that passenger numbers will double by 2040 (IATA, 2020a). Such growth in airline travel brings with it an increase in emissions, emphasizing how critical the topic of sustainability is in the industry.

2.2 Developing Long-Term Sustainability Strategies

To develop long-term sustainability strategies in the airline industry, we consider two key frameworks: the Natural Step's backcasting model and Huber's (1995, 2000) model of efficiency-consistency-sufficiency.

2.2.1 Backcasting for Sustainability

Backcasting is the process of identifying a desired future state and then taking the steps necessary to reach that state (Robinson, 1982). Unlike forecasting, backcasting does not rely on future scenarios. Backcasting is therefore a useful approach to formulating a (sustainability) strategy that works well in uncertain business contexts and overcomes path dependencies associated with forecasting. Backcasting is especially relevant for sustainability because sustainability is:

- A *complex problem* affecting many sectors and levels of society
- An issue of *externalities* that concerns *long-time horizons*
- Requires *major changes*
- A *dominant trend* (e.g., climate change, biodiversity loss) that is part of the problem

The Natural Step (2011) framework describes the A-B-C-D method for backcasting (Fig. 2). The first step (A) defines a desired sustainability future with *Awareness and visioning*. Next is (B) *Baseline mapping* of the current state to identify the gap between the future vision and where an organization is now. In step (C), organizations seek *Creative solutions* to achieve the envisioned future without any constraints. Finally, in (D) organizations *Decide on the priorities* that will help the organization move quickly towards sustainability, optimizing

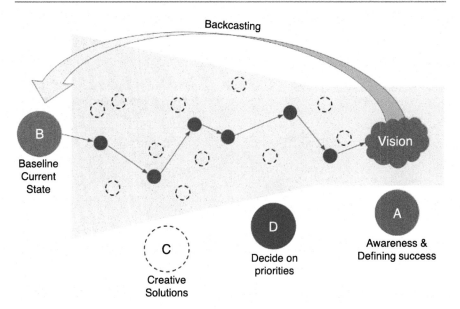

Fig. 2 The funnel metaphor and the ABCD-procedure of the FSSD. Own illustration based on Broman and Robèrt (2017, p. 21)

organizational flexibility and maximizing social, environmental, and economic returns. Due to ever-shrinking availability of natural resources, the boundaries of the natural and social systems grow ever narrower; essentially, the future can be viewed as a funnel in which unsustainable businesses risk hitting the walls of the funnel, facing loss of legitimacy, access to scarce resources, fines, legislative changes and so on (Broman & Robèrt, 2017).

2.2.2 Efficiency, Consistency, and Sufficiency for Sustainability

While the A-B-C-D backcasting method is helpful in envisioning a sustainable future without considering path dependencies and constraints, it falls short of providing concrete sustainability strategies for guidance. How, exactly, can airlines achieve sustainability? What strategies should airlines consider. To answer this question, we review Huber's (1995, 2000) grouping of three different strategies to achieve sustainability:

- *Efficiency*: a strategy of optimizing material and energy use in the production process by improving the input-to-output ratio; minimizing resource consumption and extraction via technological advances
- *Consistency*: a strategy of transformation of the whole system through a circular economy approach (closing the loop); synergistic and symbiotic flow of man-made and biological materials
- *Sufficiency*: a strategy of self-limitation by lowering production and consumption; adjusting lifestyles and consumption patterns through behavioral change

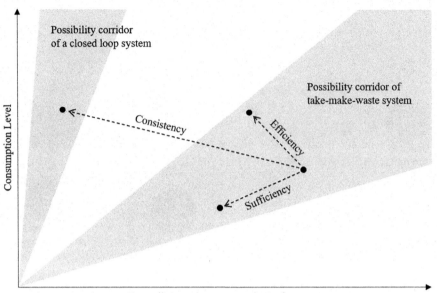

Fig. 3 Visualization of the three sustainability strategies. Own Illustration adapted from Schmidt (2008, p. 14)

The concept of these three pillars has been widely used to discuss strategies towards sustainability by organizations, policy makers, and industries (see e.g., Allievi et al., 2015; Garnett, 2014; Samadi et al., 2017; Schäpke & Rauschmayer, 2014; Schmidt, 2008). To achieve a desired future sustainability state, all three strategies are relevant for airlines. For example, the aviation industry can make improvements in engine efficiency in order to use less fossil fuel. This not only lowers emissions but also reduces costs. However, efficiency strategies are unlikely to achieve sustainability on their own, due to the so-called "rebound effect" in which improvements in efficiency lead to an even greater use of the resource that exceeds the amount saved (Figge et al., 2014). In addition, consistency strategies are required, to close the resource loop, for example, through using renewable energy sources. Finally, airlines must also build in behavioral change to reduce the overall amount of flying. This strategy might sound counterintuitive at first in an economic system designed for growth. However, by diversifying their business and corporate-level strategies, airlines gain in resilience and reduce their risk exposure to being in a carbon-intensive industry. Higher ticket prices would, for example, slow demand while raising profit margins allowing for investment into efficiency and consistency strategies (Fig. 3).

As such, the efficiency-consistency-sufficiency model offers many strategic sustainability opportunities for airlines outlined below (Table 1). As well as offering opportunities, each strategic approach has limitations that airlines need to address. For example, *efficiency strategies* represent important cost savings for airlines, with

Table 1 Three sustainability strategy approaches for airlines

Strategy	Definition	Meaning for airlines	Example approaches	Limitations/ drawbacks
Efficiency	Optimize resource use	Continuous improvement of technologies and operations	• "Single European Sky" could reduce fossil fuel use by 10% (Kettunen et al., 2005) • Optimizing air traffic routes could reduce fossil fuel use by 6–12% (IPCC, 1999)	• Slow progress • Fuel efficiency unlikely to exceed 2% improvement (ICAO, 2016) • Technological and operational improvements only consist of 1.4% annual fuel efficiency progress (ICAO, 2019) • Efficiency approaches will not outweigh booming demand (ETC, 2018b; IATA, 2019a)
Consistency	Transforming the system	Replacing current technologies with ones that close the loop	• Alternative engines (electric, hydrogen) • Sustainable aviation fuels potentially reduce CO_2 emissions by up to 80% (ATAG, 2017; ETC, 2018a, b; IATA, 2019a; ICAO, 2019)	• Alternative engines not viable for long-haul flights by 2050 goals (ATAG, 2020) • Biofuels compete with food production (ICAO, 2019), 33 million hectares needed vs. 8 million hectares for synthetic fuel (Transport & Environment, 2017)
Sufficiency	Behavioral change	Reducing number of flights and flight demand	• France's regulation to ban flight routes accessible by train within 2.5 h • Collaborations with high-speed rail, such as SWISS	• Approach faces strong criticism and resistance from industry and regulators • Requires global coordination to level the competitive field • 80% of aviation's emissions globally are caused by long-haul flights (ATAG, 2020), which are more difficult to substitute

a direct financial benefit as well as having some environmental benefits. So far, the most progress has been made in these strategies. However, if implemented without other strategies, efficiency strategies fall drastically short of reaching the climate goals set out in the Paris Agreement and thus are only a temporary solution while the aviation industry pushes forward with more transformational strategies. As such, consistency and sufficiency strategies are needed to decarbonize the aviation industry.

Transformational approaches like *consistency strategies* face their own challenges as alternative engines, for example, are not currently viable for long-haul flights, and some alternative fuels like biofuel compete with food production. Development of synthetic fuels is therefore seen as the most viable option in the coming years, as they can be blended with conventional fuels and used in current aircraft and likely to replace petroleum-based fuel completely (IATA, 2019a; ICAO, 2016). However, synthetic fuels are currently only available in small quantities and two to seven times as expensive as conventional kerosene; scaling their production requires large investments (BAZL, 2019; Hong et al., 2019). Since fuel accounts for roughly one third of airlines' total expenses, the high price of SAF poses a significant barrier. The fierce price competition between airlines makes voluntarily heightened costs or large investments difficult. ICAO argues that the complete replacement of fossil fuels with sustainable alternative fuels is physically possible, but "would require substantial policy support" (ICAO, 2019, p. 20). If such support were to be provided, the price of synthetic fuel would quickly decline due to economies of scale and experience curve effects (ETC, 2018b). Adding a carbon tax or blending quota policy would make synthetic fuel pricing more attractive in comparison to conventional fuels. Support for SAF as the main instrument to achieve sustainability in the industry is evident as airlines have signed letters of intent to buy 6 billion liters of SAF, and airports have agreed on providing the necessary infrastructure (IATA, 2019b, 2020b). However, 6 billion liters only represents 1.4% of total fuel consumption in the airline industry in 2019, which was 430 billion liters (Statista, 2020).

In sum, although synthetic fuels are not completely GHG neutral (see Chap. "Technology Assessment for Sustainable Aviation"), they are the most likely sustainable technology alternative for commercial long-haul air travel. The fact that the technology already exists and sustainable fuel can be used in today's aircraft makes synthetic fuels a promising option in decarbonizing aviation. Implementing this approach depends on commitment and coordinated efforts by policy and other stakeholders.

Sufficiency strategies such as reducing demand for flying remain the elephant in the proverbial airline cabin. Studies have shown that a reduction in demand is necessary to keep global warming below 2 °C (Bows-Larkin et al., 2016). However, industry players and regulators are highly critical of discussions around reducing demand, arguing that such expectations are unrealistic both from a consumer and a supplier perspective. Our global world is highly mobile and interconnected, and a complete elimination of air travel is unlikely. However, a reduction in air travel is necessary as we combat climate change, at least until there are suitable alternatives to kerosene-powered flying.

Common approaches to limit air travel are policy measures like ticket taxes (see Chap. "The Role of Public Policy"). Higher prices lead to less demand, while the tax income can be used to commercialize synthetic fuels. A concern raised in this context is that higher ticket prices discriminate against lower income classes. However, from a global perspective, it is more discriminating to *not* take action against climate change, as low-income classes will be hit the hardest by its consequences.

Next to policy measures, air travel can be replaced by alternative services that make flights redundant. This is especially a possibility for business travel that can be substituted with videoconferencing to conduct meetings. The COVID-19 pandemic has shown that videoconferencing is a viable alternative to air travel, and many businesses see the benefits of traveling less. Substituting leisure travel is more difficult. However, even here, air travel can be reduced by supporting train travel at local level. The COVID-19 crisis has also shown customers that local and regional tourism meets leisure travel needs (Kirig, n.d.). Developing local tourism services is a strategic opportunity for airlines.

A successful example of a modal shift to more sustainable forms of travel is the high-speed train TGV in France to substitute short-haul flights (Bachman et al., 2018). Customers are increasingly opting for trains on shorter routes. In Germany, air passenger numbers were lower in 2018 than the previous year (ADV, 2020), while the number of long-distance rail passengers is higher than ever before (DB, 2020). High-speed rail is growing steadily across Europe (European Court of Auditors, 2018). This represents a strategic opportunity for airlines to switch to lower emission transportation modes on short-haul routes, for example, through partnerships with train services, especially if airlines continue to provide support services to passengers in a holistic "flight system" that takes care of delays, rebookings, and luggage transport.

A limitation of focusing on replacing short-haul flights is that these flights only cause 20% of aviation's emissions globally (ATAG, 2020). The Energy Transitions Committee estimates that "if a third of all short-haul passenger traffic was shifted to rail by mid-century, only about 10% of total emissions from aviation would be eliminated" (ETC, 2018b). Nevertheless, a 10% reduction of emissions is still valuable. A modal shift to rail depends on factors like travel time, price, reliability and comfort. Travel time and price are particularly important for business travelers (UIC, 2018). Investment would also be needed into an international high-speed rail network. To date, such investments have been slow to progress. For example in Europe, the EU's goal of tripling high-speed rail kilometers by 2030 is not likely to be achieved (European Court of Auditors, 2018) mainly due to lack of sufficient international cooperation and prioritization.

Initiatives like the 2021 "European Year of Rail" may result in shifting demand from air to rail on short-haul routes (European Commission, n.d.). The Swiss government likewise has encouraged its national rail system to link up to the European high-speed rail network (UVEK, 2018). Swiss and Austrian rail companies are also working to improve night train services (Rensch, 2019) which have been increasing in demand. Such initiatives can support Rail&Fly programs where customers can combine both mobility modes. Globally, one third of all short-haul traffic could be shifted to rail by 2050 (ETC, 2018a). For Europe, the potential is higher, since the market share of short-haul flights is above the global average[1] (Boeing, 2019) and even higher in Switzerland. Two-thirds of all passengers (67.7%) departing from a Swiss airport started their journey in Switzerland with an ultimate destination in Europe (BFS, 2019). A stronger European high-speed rail network rail network could therefore lead to a significant reduction in the number of short-haul flights from Switzerland.

Disruptive technologies like maglev (magnetic levitation) trains have high potential to reduce emissions from travel. Powered by magnetic repulsion, maglev trains lift from the ground after reaching 150 km/h, eliminating friction and allowing extremely high speeds with the current record in Japan at 603 km/h (Japan Rail Pass, 2019; UIC, 2018). This speed approaches average passenger aircraft speeds of around 900 km/h (Morris, 2017). Maglev investments are particularly prominent in China and other parts of Asia (Bachman et al., 2018; UIC, 2018). Other technologies like the Hyperloop in the USA could likewise "seriously change the current nature of air travel" (IATA, 2019a, p. 43), presenting another strategic opportunity for mobility companies like airlines.

To conclude, sufficiency strategies represent a promising approach to reducing emissions, although mostly to replace short-haul flights. Airlines pursuing these strategies need to partner with governments and rail companies to overcome barriers, but airlines' competencies in service are a source of competitive advantage. For long-haul flights, online conferencing offers a substitute for business travelers, as well as partnering with local and regional rail for end destinations of leisure travelers.

3 Case Study: Future Sustainability Strategy for SWISS

To assess the viability of the efficiency, consistency, and sufficiency strategies for an airline, the first author of this chapter conducted a case study of SWISS, or Swiss International Airlines AG, the Swiss national carrier that has been part of the Lufthansa group since acquisition was completed in 2007. The study aimed at evaluating the current environmental strategy of SWISS and develop future strategies to achieve full environmental sustainability, i.e., to be in accordance with mitigating climate change to 2 °C. To achieve this goal, the researcher conducted three in-depth interviews with SWISS employees and eight of SWISS's stakeholders. When selecting the interview partners, care was taken to ensure that several different stakeholder groups of SWISS (e.g., policy, investors, research)

---

[1] Short-haul market share in Europe is 42%. This is the third-highest share globally, after Southeast Asia (62%) and South Asia (57%). Europe is followed by Latin America (35%) and North America (32%). In all other regions, short-haul market shares are below 30%. The global average of short-haul market share is 33%.

were represented. Also, two interview partners were selected specifically because of their public and strong advocacy for environmental action. These semi-structured interviews were conducted between February 26, 2020, and April 3, 2020, and formed the primary basis for the study. Implications from the COVID-19 pandemic were not systematically discussed and excluded from analysis. Also, additional desk research was conducted to complement or cross-check the primary data from the interviews. The desk research focused on relevant reports by the industry (e.g., by IATA, ICAO, Lufthansa) or other larger institutions (e.g., by European Commission, IPCC, the International Union of Railways).

3.1   Opportunities and Threats for SWISS in Three Sustainability Strategies

One goal of the case study was to identify strategic sustainability opportunities for SWISS and facilitation factors, assessed against the barriers or impeding factors. In general, interviewees stated that SWISS has many strategic opportunities for sustainability due to high public awareness of sustainability in Europe, and the importance of sustainability strategies for strategic and financial viability of the airline, as well as its investors. The industry also acknowledges that sustainability is important and efforts like CORSIA support a transition to a more sustainable industry. In contrast, general impeding factors include the short amount of time left to achieve climate goals, in combination with regulatory uncertainty and a general short-term horizon of shareholders. In addition, sustainability is not seen as a priority in all countries, and interviewees believed sustainability might fade from public attention. A lack of coordination on taxes was seen as another barrier to decarbonization of the industry.

The analysis of SWISS's opportunities and threats in the three types of sustainability strategy (Table 2) showed that recurring barriers in the decarbonization of aviation are the fierce price competition (which benefits efficiency but hinders consistency or sufficiency strategies) as well as the dependence on external stakeholders such as policy makers and governments to play an active role, a similar point made by ICAO (2019). Investments in research and development are needed, as well as policies supporting the deployment of sustainable fuel.

Despite these difficulties, there are also a lot of opportunities. This shows that airlines can be proactive about engaging in sustainability strategies rather than passing off all responsibility to decarbonize the sector to governments and international institutions. The urgency of the climate crisis requires all actors to proactively exploit all possible opportunities to become more sustainable. This not only benefits the natural ecosystem, but is also linked to competitive advantage for airlines, which can drive transformation across the entire sector.

As outlined earlier, the mitigation potential of efficiency measures to lower emissions is rather limited for airlines like SWISS but should nevertheless be exploited since it brings financial gains. The core focus of airlines' sustainability strategies should be on consistency and sufficiency.

Table 2 Opportunities and threats for SWISS becoming sustainable

	Opportunities	Threats
Efficiency	– Large efficiency potential in airspace design – Fierce price competition encourages efficiency measures	– Several political barriers for international collaboration – Mitigation potential of operational and engine efficiency measures soon exhausted
Consistency	– Electric or hydrogen aircraft could become feasible on short-haul flights by 2050 – Synthetic fuels as a very promising solution, especially for long-haul flights, due to their technical feasibility and applicability – Revenues of ticket taxes could support synthetic fuels	– Viable alternative engine technologies for long-haul not realistic by 2050 – Long lifecycle of aircrafts – Synthetic fuels are not completely GHG neutral – Synthetic fuels currently only available in small quantities and very expensive – High price sensitivity of customers (goes along with low willingness to pay for sustainability) – Fierce price competition and economic system designed for growth impedes new technologies requiring large investments like synthetic fuel – Decarbonization greatly depends on policy & many other stakeholders – Immense demand growth of air travel expected, which means more sustainable fuel is needed
Sufficiency	– Improved and widespread videoconferencing could reduce business travel – Customers recognizing trains as a substitute (indicated by increased demand for rail travel) – High-speed rail can be a comparable short-haul service, but with significantly lower emissions – High market share of short-haul flights in Europe – Increasing effort of governments and rail companies to improve train network – Ticket taxes being introduced could throttle demand growth	– Raising ticket prices and reduction of mobility is controversial – Customers combining short-haul and long-haul reject multimodal traveling – Lacking quality of European rail network – Overall, emission reduction potential of a modal shift is limited (as it only applies to short-haul) – No mobility alternatives for long-haul flights – Global regulations would be needed to raise prices and throttle growth, but are hard to implement

Own illustration

3.2 Designing a Visionary Sustainability Strategy for SWISS

A second goal of the case study was to envision a substantive sustainability strategy for SWISS that aims to be in accordance with the goal of mitigating climate change to below 2 °C. To design a more visionary sustainability strategy, using the A-B-C-D framework, focusing on steps C and D, an exploratory ideation approach was used. As each airline's strategic context and competencies are unique, the design adopted here is not meant to represent a "best" strategy but rather serve as an illustration of how an airline might pursue a visionary sustainability strategy. Here, we outline two (out of many) possible approaches. In the first scenario, we visualize an emphasis on a consistency strategy with a focus on replacing fossil fuels. In the second scenario, we visualize a sufficiency strategy with a focus on SWISS becoming a multimodal mobility provider.

3.2.1 SWISS Scenario 1: Consistency Strategy—Replacing Fossil Fuels

In the first scenario, an assumption is made that little action is taken by policy makers and consumers to decarbonize the aviation sector, and technological transformation to alternative aircrafts is minimal. As such, this scenario is characterized by little proactive government support for synthetic fuels, no restrictions in air travel and steady demand growth, sinking climate change awareness among the public, little willingness of consumers to switch to trains, and little progress of the European high-speed rail network. Airlines keep to their core business as an aviation mobility provider.

In this scenario (Fig. 4), a recommended approach would be to fully focus on consistency strategy. The goal is to replace fossil fuels with synthetic fuels and offset the remaining GHG emissions with a few but highly effective compensation projects. To achieve this goal, airlines collaborate with synthetic fuel producers and advocate publicly for supporting synthetic fuels. To receive financial support, airlines could apply for special aviation funds (e.g., the "Spezialfinanzierung Luftverkehr" in Switzerland) to support this initiative and help kickstart it. Towards policy makers, airlines could advocate for compulsory blending quotes of fuels and ticket taxes to support the development of synthetic fuels. One option could also be for CORSIA and ETS fees to flow into synthetic fuel subsidies and promote their use for compensation projects instead of generic climate projects. Also, airlines could engage in campaigns to raise consumer awareness and promote voluntary compensation by creating incentive and reward systems.

In a more disruptive consistency strategy, airlines would engage in a backward vertical integration strategy to acquire synthetic fuel production companies. In doing so, airlines would be entering a growing market and secure access to synthetic fuel, making them less dependent on external stakeholders to achieve their sustainability vision and securing a financial benefit. A consistency strategy thereby becomes centrally embedded in airlines' business models.

Naturally, a focusing on synthetic fuels raises certain risks, as it is based on several assumptions: (1) synthetic fuel production can be scaled until 2050, (2) it can be powered solely by renewable energy and (3) its price will become competitive with fossil fuels. For all these assumptions, there are barriers and dependencies: What if the decrease in price is smaller than predicted? What if there are no subsidies from the state or other policy mechanisms? What if the share of customers compensating declines instead of increases? What if the industry loses interest in the technology and the airline remains the only airline investing in synthetic fuels?

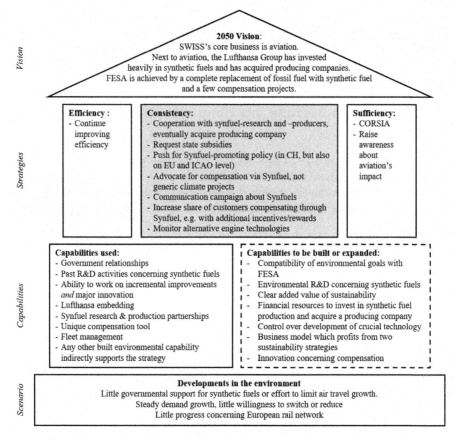

Fig. 4 Strategy Overview "Green Status Quo." Own illustration. Note. Shaded areas represent efforts on which the strategy focuses. FESA stands for Full Environmental Sustainability for Airlines

These questions all point to the main flaw of this strategy: risk is not diversified. Instead, the strategy focuses almost exclusively on one technology and does not pursue other measures to significantly reduce emissions.

3.2.2 SWISS Scenario 2: Sufficiency Strategy—Multimodal Mobility

In the second scenario, an assumption is made that governments show considerable support for synthetic fuels, bring in policies that limit air travel growth, a slowing of consumer demand in long-haul flights and shifting to trains for short-haul routes, and a strengthening of the European high-speed rail network. In this scenario, airlines could rethink their core business and consider becoming multimodal mobility players by extending their services beyond flights into rail travel. The central purpose of airlines is no longer to focus on flight routes but rather to move customers from location A to B and become end-to-end mobility providers. Their services

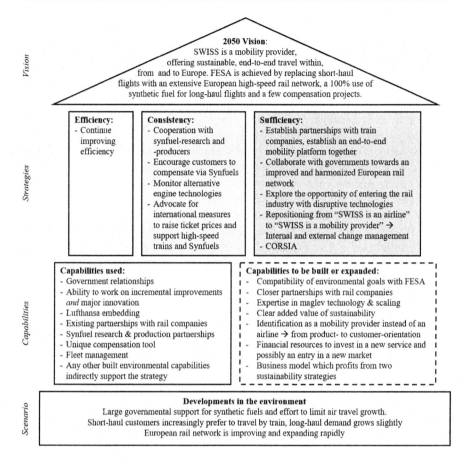

Fig. 5 Strategy Overview "Multimodal Mobility." Own Illustration. Note. Shaded areas represent efforts on which the strategy focuses. FESA stands for Full Environmental Sustainability for Airlines

include long-haul flights powered by synthetic fuel and high-speed rail lines for short-haul routes.

Following the Mobility-as-a-Service approach (Jittrapirom et al., 2017), customers would be able to book seamless, international door-to-door trips in one session. While booking the trip, voluntary compensation is automatically suggested, and customers receive one ticket that is valid for the whole trip. Luggage is taken care of from start to completion of the customer's journey. There is only one contact point for the whole trip: if there is any problem (e.g., a missed connection), customers can reach out to the service provider, no matter where they currently are on their journey. Airline's loyalty programs, for example Lufthansa's Miles&More, are integrated into rail travel. Services could further expand to other mobility modes such as public busses, taxis, bike sharing, etc. (Fig. 5).

Such a strategy corresponds with up-and-coming trends towards customer-oriented, multimodal mobility (Rauch, 2020). The think tank Zukunftsinstitut predicts that mobility providers can no longer solely focus on their core service, but need to shift to a customer-centered perspective and offer need-specific, end-to-end services (Huber et al., 2011). This strategy also counters the risk of decline or stagnation in air travel demand. Routes like London-Paris or Paris-Zurich have shown that rail is a real alternative short-haul flights. By engaging in a multimodal sufficiency strategy, airlines continue to benefit from short-distance travel instead of losing customers to rail companies. In addition, extending services to rail anchors sustainability more deeply in an airline's business model, meaning that airlines profit not only from efficiency but also from sufficiency approaches.

Compared to the first scenario, the sufficiency strategy in this second scenario diversifies the risk exposure of airlines. However, a major weakness of this sufficiency strategy is that it does not lower emissions to the needed degree: short-haul flights represent only 20% of aviation's emissions globally, and long-haul flights for the remaining 80% (ATAG, 2020). This means that airlines still need to solve the problem of decarbonizing long-haul routes. Airlines would consider complementary strategies as they arise, such as disruptive innovation in alternative fuel aircrafts (electric, hydrogen) or other new inventions.

A second weakness of the sufficiency strategy is that international high-speed rail networks still need a lot of infrastructure development for connectivity and reliability. The slow progress on this front makes a competitive network unlikely in the coming decades. In Europe, it takes on average 16 years until a new high-speed line begins to operate (European Court of Auditors, 2018). Moreover, building a railway network also creates emissions. However, in the long run, such investment is more likely to address climate change than continued growth in aviation.

4 Short-Term Sustainability Strategic Options for Airline

It is clear that airlines need to invest in new capabilities and competencies in the long-term because efficiency will not be enough to mitigate climate change. Rather airlines need to consider sufficiency and consistency strategies to make a meaningful impact, such as reducing flights and exploring disruptive technologies. However, the topic of climate change is so urgent that airlines should act now. What can airlines do in the short term to mitigate GHG emissions? Using a comprehensive review of literature on existing sustainability strategies of airlines, complemented by interviews with the airline and climate change experts using the Mayring method (Mayring, 2020), we identify viable, short-term climate change strategies for airlines.

While short-term climate mitigation actions are unlikely to have a very large impact on reducing GHG emissions, they nevertheless can be meaningful because incremental changes add up. By embarking on the path to mitigate climate change through strategic action in the short term, airlines benefit in two ways. First, airlines engrain a cultural shift in their companies and with their stakeholders that establishes a new mindset on sustainability. Second, airlines invest in capabilities that can lead to future competencies, securing their competitive position in the sector while maintaining legitimacy. Here we use the upside-down pyramid approach (Chap. "Introducing Sustainable Aviation Strategies") to prioritize strategies with higher climate mitigation impact first: avoid, reduce, replace/switch, offset.

4.1 Avoiding Emission Strategies

To avoid GHG emissions in the short term, engaging in *strategic partnerships with high-speed rail* companies and/or electric bus routes are a viable strategy for airlines. Such partnerships do not require long lead times, and airline companies can leverage their competence for service and one-stop ticket booking in such partnerships. This would allow airlines to reduce up to 20% of their emissions due to avoiding flights on short-haul routes (ATAG, 2020). Examples include partnerships with Emirates and Deutsche Bahn and other European countries (Emirates, 2020a, b), which has a Rail&Fly ticket that links people from major European cities to their final destination. SWISS and Lufthansa similar embark in partnerships with SBB and Deutsche Bahn (Lufthansa Group, 2020a; SWISS, 2020).

4.2 Reducing Emission Strategies

The next step is *reducing* emissions, both through technological efficiency and route optimization gains. Ensuring *technological efficiency* such as airplane design and engine fuel use is another viable short-term strategy, especially for airlines with older infrastructure. New generations of energy-efficient aircraft reduce fuel use by 25% (Klimaschutzportal, 2020) as they are lighter and more efficient. However, for airlines that have already upgraded their infrastructure, technological efficiency solutions are likely to gain only marginal further increases (see Chap. "Perceptions of Flight Shame and Consumer Segments in Switzerland"). Moreover, fleet renewal requires high investment costs, and the recent pandemic may create problems with liquidity among airlines, as well as delays in orders and deliveries. Ground transport can also be switched to renewable and electric energy vehicles. The Lufthansa Group plans to achieve CO_2 neutrality "on the ground" by 2030 in its key markets. From a policy perspective, airlines can engage with policy makers and governing bodies to push for measures that avoid ticket price distortions (like CORSIA), and also the *optimization* of flight routes such as the Single European Sky, which can save up to 10% of GHG emissions (Kettunen et al., 2005) and landing routes which can save 6–12% (IPCC, 1999). Such approaches are explored in more depth in Sect. 3.3 in Chap. "Perceptions of Flight Shame and Consumer Segments in Switzerland."

Another way in which airlines can reduce GHG emissions in the short term is through *closing the supply chain loop*, such as by managing waste produced on flights. In 2017, flights produced 5.7 million tons of waste through catering, packaging, and other on-board services (Aviation Benefits, 2020; IATA, 2020d), or 1.43

kilograms of waste per passenger (IATA, 2020d). Some 23% of waste is untouched food which is directly destroyed after landing rather than recycled due to biohazard regulation. Similarly, hygiene rules lead to increased use of plastic. Qantas has managed to reduce on-board waste by passing on 18,000 leftover meals a month to aid organizations (Aviation Benefits, 2020) and has goals to decrease its waste by 75% by the end of 2021. Airlines like the Lufthansa Group have banned disposable plastic using recycled beverage cups, and Etihad Airways plans to reduce single-use plastic by 80% by the end of 2021.

4.3 Replacing/Switching Emission Strategies

Next, airlines can replace/switch to "better" emissions like ***shifting to SAFs***, which can be used in the current engine technology and infrastructure. This is valuable as overall GHG emissions of SAFs are an estimated 80% lower than kerosene (ATAG, 2017; IATA, 2019b). As discussed earlier, synthetic fuels are a viable long-term strategy but not yet commercially available. In the short term, therefore, airlines can opt for biofuels while engaging in longer term strategies to scale up and commercialize synthetic fuels. However, in the long term, this strategy is not viable because of the competition of biofuels with food (ICAO, 2019).

4.4 Offsetting Emission Strategies

As a last resort, offsetting or compensation can be considered by airlines. This is more a temporary tactic than a strategic solution because of the problems and limitations with offsetting compensation schemes (Boon et al., 2006; Gössling et al., 2007). Nevertheless, airlines can use ***nudging techniques*** to encourage customers to consider offsetting when booking their flights by changing the default options. Such an approach, which is based on the "status quo bias" (Thaler & Sunstein, 2008, p. 7), is likely to increase the proportion of customers who will compensate for their flight emissions. Another option is ***incentives*** like a glass of wine on the flight as a reward for offsetting. As discussed in Chap. "Introducing Sustainable Aviation Strategies," offsetting should be considered only after all other climate mitigation steps have been implemented as it does not meaningfully address climate change.

4.5 Developing a Climate Change Culture to Build Future Competencies

In creating a cultural shift to embed a sustainability mindset into their business, airlines ready themselves for long-term climate change action by building up necessary capabilities that can lead to competencies to mitigate climate change. The first step includes integrating sustainability into the core strategy and business processes of the company. By continually scanning the ever-changing business context, challenging old assumptions, understanding the implications, recognizing opportunities and transforming their organizations, companies build dynamic capabilities through sensing, seizing, and transforming (Linden, 2021). Thus, companies always need to keep an eye on the long-term goal, through a process like backcasting. In doing so, companies can build sustainability-enabling capabilities that will help them to transform their organizational culture and business model (Bertels et al., 2010; Doppelt, 2017; Galpin et al., 2015; Mayer et al., 2014; Walls, 2018; Walls et al., 2011).

Sustainability-enabling capabilities include taking a long-term perspective (vision), top leadership commitment, embedding sustainability into the core strategy of the company, developing science-based targets that can be tracked systematically and transparently reported (e.g., the GHG Protocol, GRI), sustainability sensitizing and training of all employees and stakeholders, sustainability management systems such as sustainable procurement practices, an adaptive and flexible culture that is open to change, an incremental and breakthrough innovation approach, incentive schemes for managers and employees that tie job performance to sustainability performance, and viewing sustainability as an opportunity rather than a constraint.

5 Conclusion

The aviation industry faces much uncertainty, operating in a volatile and complex business context with an urgent need to mitigate its impact on climate change. Airlines face pressure from investors, consumers, activists, regulatory bodies, peers, and other stakeholder groups to transform their business models towards sustainability. Airlines are not ignorant of the need to reduce emissions. IATA has long been involved in designing a comprehensive strategy to reduce the climate impact of airlines, which is followed by many airlines around the globe. However, the climate goals set by IATA are not compatible with the Paris Agreement to mitigate climate change to below 2 °C. As such, airlines need to do much more. The good news is that engaging in climate change strategies provides strategic opportunities for airlines both in the short- and long term, and secures their legitimacy and competitiveness.

In this chapter, we evaluate the efficiency-consistency-sufficiency model, sketching two possible future scenarios. Airlines are already engaging in efficiency strategies such as optimizing routes which have reduced fuel consumption by a third. However, future efficiency gains are likely to be incremental rather than radical and in the long term, airlines need to turn to more impactful strategies. Our analysis shows that airlines can significantly reduce emissions through a consistency strategy, which focuses on synthetic fuels as an alternative to kerosene. While availability and prices of synthetic fuels currently represent a barrier, by partnering and investing in this market, the industry could scale up production as an almost net neutral alternative to fossil fuels without requiring major changes in engine technology. Reducing the demand for air travel, via a sufficiency approach, is a third viable strategy that focuses on substituting flights with high-speed rail such as maglev, especially on shorter routes. However, this strategy depends heavily on infrastructure investments across regions. Other sufficiency strategies to reduce demand, such as through higher ticket pricing, require a policy-driven approach to level the competitive playing field.

In the short term, airlines can reduce emissions by applying the upside-down pyramid model to avoid-reduce-switch-offset, based on insight of the industry and experts. Many of these approaches are precursors to long-term strategies to reduce emissions. By engaging in these short-term strategies now, airlines embed a sustainability mindset into the culture of their organizations which leads to the development of dynamic capabilities to ensure airlines build climate mitigation competencies for the future.

To conclude, the low-hanging fruits of climate mitigation have already been picked by many airlines, and a more fundamental transformation of airlines' business models is needed in order to achieve net carbon neutrality by 2050. This means replacing fossil fuels, which are the largest source of (direct, scope 1) emissions for airlines and reducing the amount of flying. Such a change may seem radical but it is necessary and possible, and airlines can survive and compete if they take appropriate action.

References

ADV. (2020, December). *ADV-Monatsstatistik.*' Retrieved April 11, 2020, from https://www.adv.aero/wp-content/uploads/2016/02/12.2019-ADV-Monatsstatistik.pdf

Aerotelegraph. (2020). *Paris Hilft Air France - Wenn Airline Grüner Wird.*

Airbus. (2020). *Airbus reveals new zero-emission concept aircraft.*

Allievi, F., Vinnari, M., & Luukkanen, J. (2015). Meat consumption and production – Analysis of efficiency, sufficiency and consistency of global trends. *Journal of Cleaner Production, 92*, 142–151. https://doi.org/10.1016/j.jclepro.2014.12.075

ATAG. (2017). *Beginners guide to SAF.* Retrieved March 10, 2020, from https://aviationbenefits.org/media/166152/beginners-guide-to-saf_web.pdf

ATAG. (2020). *FACTS & FIGURES.* Retrieved March 16, 2020, from https://www.atag.org/facts-figures.html

Aviation Benefits. (2020). *Circular economy.* Retrieved April 10, 2020, from https://aviationbenefits.org/environmental-efficiency/circular-economy/

Bachman, J., Fan, R., & Cannon, C. (2018, January 10). Watch out, airlines. High speed rail now rivals flying on key routes. *Bloomberg.*

Bansal, P., & Clelland, I. (2004). Talking trash: Legitimacy, impression management, and unsystematic risk in the context of the natural environment. *Academy of Management Journal, 47*(1), 93–103. https://doi.org/10.2307/20159562

BAZL. (2019). *Alternative Treibstoffe in Der Schweizer Zivilluftfahrt.* Retrieved December 9, 2020, from https://www.bazl.admin.ch/dam/bazl/fr/dokumente/Politik/Umwelt/Umweltmassnahmen/Alternative_Treibstoffe_in_der_Schweizer_Zivilluftfahrt.pdf.download.pdf/Alternative_Treibstoffe_in_der_Schweizer_Zivilluftfahrt.pdf

Berns, M., Townend, A., Khayat, Z., Balagopal, B., Reeves, M., Hopkins, M., & Kruschwitz, N. (Eds.). (2009). *The business of sustainability. Imperatives, advantages, and actions.* Boston Consulting Group.

Bertels, S., Papania, L., & Papania, D.. (2010). *Embedding sustainability in organizational culture. A systematic review of the body of knowledge.* Network for Business Sustainability.

BFS. (2019). *Schweizerische Zivilluftfahrtstatistik 2018 - 5. Passagiere*. Retrieved March 24, 2020, from https://www.bfs.admin.ch/bfsstatic/dam/assets/9386714/master

Boeing. (2019). *Commercial Market Outlook 2019 – 2038. Market analysis*.

Boon, B. H., Schroten, A., & Kampman, B. (2006). Compensation schemes for air transport. *Paper Written for the E-CLAT Climate Change and Tourism Conference in the Netherlands*.

Bows-Larkin, A., Mander, S. L., Traut, M. B., Anderson, K. L., & Ruth Wood, F. (2016). Aviation and climate change - the continuing challenge. In R. Blockley & W. Shyy (Eds.), *Encyclopedia of aerospace engineering* (pp. 1–11). Wiley.

British Airways. (2020). *Corporate responsibility*. Retrieved March 20, 2020, from https://www.britishairways.com/en-gb/information/about-ba/csr/corporateresponsibility

Broman, G. I., & Robèrt, K.-H. (2017). A framework for strategic sustainable development. *Journal of Cleaner Production, 140*, 17–31. https://doi.org/10.1016/j.jclepro.2015.10.121

Compensaid. (2020). *Wählen Sie Ihre Kompensation*. Retrieved July 10, 2020, from https://compensaid.com/contribute/split?flights=06f6a147-615a-4c19-a1ba-79eb34227a70

DB. (2020). *DB 2019: Erstmals Über 150 Millionen Reisende Im Fernverkehr • DB Regio Mit Erfolgreicher Trendwende • Investitionsoffensive – Vor Allem Bei DB Netze – Ausgebaut*. Retrieved April 11, 2020, from https://www.deutschebahn.com/de/presse/pressestart_zentrales_uebersicht/DB-2019-Erstmals-ueber-150-Millionen-Reisende-im-Fernverkehr-DB-Regio-mit-erfolgreicher-Trendwende-Investitionsoffensive-vor-allem-bei-DB-Netze-ausgebaut%2D%2D5049890?contentId=1204030

Delta. (2020). *Delta commits $1 billion to become first carbon neutral airline globally*. Retrieved January 16, 2021, from https://news.delta.com/delta-commits-1-billion-become-first-carbon-neutral-airline-globally

Doppelt, B. (2017). *Leading change toward sustainability: A change-management guide for business, government and civil society* (2nd ed.). Routledge.

easyJet. (2019). *Annual Report and Accounts 2019*.

easyJet. (2020). *Annual Report 2019*. Retrieved March 16, 2020, from https://corporate.easyjet.com/~/media/Files/E/Easyjet/pdf/investors/resultscentre/2019/eas040-annual-report-2019-web.pdf

Eccles, R. G., Ioannou, I., & Serafeim, G. (2014). The impact of corporate sustainability on organizational processes and performance. *Management Science, 60*(11), 2835–2857. https://doi.org/10.1287/mnsc.2014.1984

Emirates. (2020a). *Codeshare partnership with Deutsche Bahn*. Retrieved August 5, 2020, from https://www.emirates.com/de/english/travel-partners/deutsche-bahn/

Emirates. (2020b). *Travel partners*. Retrieved August 5, 2020, from https://www.emirates.com/de/english/travel-partners/

Emirates. (2021). *How we fly our planes*. Retrieved May 20, 2021, from https://www.emirates.com/qa/english/about/emvironment/how_we_fly.aspx

ETC. (2018a). *Mission possible: Reaching net-zero carbon emissions from harder-to-abate sectors by mid-century*. Retrieved June 2, 2020, from http://www.energy-transitions.org/sites/default/files/ETC_MissionPossible_FullReport.pdf

ETC. (2018b). *Mission possible: Reaching net-zero carbon emissions from harder-to-abate sectors by mid-century. Sectoral focus aviation*. Retrieved June 2, 2020, from http://www.energy-transitions.org/sites/default/files/ETC%20sectoral%20focus%20-%20Aviation_final.pdf

European Commission. (n.d.). *Reducing emissions from aviation. Policy*. Retrieved May 20, 2020, from https://ec.europa.eu/clima/policies/transport/aviation_en

European Court of Auditors. (2018). *A European high-speed rail network: Not a reality but an ineffective patchwork*. Retrieved May 24, 2020, from https://www.eca.europa.eu/Lists/ECADocuments/SR18_19/SR_HIGH_SPEED_RAIL_EN.pdf

Figge, F., Young, W., & Barkemeyer, R. (2014). Sufficiency or efficiency to achieve lower resource consumption and emissions? The role of the rebound effect. *Journal of Cleaner Production, 69*, 216–224. https://doi.org/10.1016/j.jclepro.2014.01.031

Finnair. (2020). *Sustainability report 2019*.

Galpin, T., Lee Whitttington, J., & Bell, G. (2015). Is your sustainability strategy sustainable? Creating a culture of sustainability. *Corporate Governance: The International Journal of Business in Society, 15*(1), 1–17. https://doi.org/10.1108/CG-01-2013-0004

Garnett, T. (2014). Three perspectives on sustainable food security: Efficiency, demand restraint, food system transformation. What role for life cycle assessment? *Journal of Cleaner Production, 73*, 10–18. https://doi.org/10.1016/j.jclepro.2013.07.045

Gössling, S., Broderick, J., Upham, P., Ceron, J.-P., Dubois, G., Peeters, P., & Strasdas, W. (2007). Voluntary carbon offsetting schemes for aviation: Efficiency, credibility and sustainable tourism. *Journal of Sustainable Tourism, 15*(3), 223–248. https://doi.org/10.2167/jost758.0

Hart, S. L. (1995). A natural-resource-based view of the firm. *Academy of Management Review, 20*(4), 986–1014. https://doi.org/10.5465/amr.1995.9512280033

Hong, Y., Cui, H., Dai, J., & Ge, Q. (2019). Estimating the cost of biofuel use to mitigate international air transport emissions: A case study in Palau and Seychelles. *Sustainability, 11*(13), 3545. https://doi.org/10.3390/su11133545

Huber, J. (1995). Nachhaltige Entwicklung Durch Suffizienz, Effizienz Und Konsistenz. In P. Fritz, J. Huber, & H. W. Levi (Eds.), *Nachhaltigkeit in naturwissenschaftlicher und sozialwissenschaftlicher Perspektive* (pp. 31–46). Hirzel, Wissenschaftliche Verlagsgesellschaft.

Huber, J. (2000). Towards industrial ecology: Sustainable development as a concept of ecological modernization. *Journal of Environmental Policy & Planning, 2*(4), 269–285. https://doi.org/10.1080/714038561

Huber, T., Rauch, C., & Volk, S. (2011). *Die Zukunft der Mobilität - 2030: das Zeitalter der Managed Mobilität. edited by Zukunftsinstitut.* Zukunfts-Inst.

IATA. (2018). *Climate change & CORSIA.* Retrieved March 24, 2020, from https://www.iata.org/contentassets/c4f9f0450212472b96dac114a06cc4fa/fact-sheet-climate-change.pdf

IATA. (2019a). *Aircraft technology roadmap to 2050.* Retrieved June 3, 2020, from https://www.iata.org/contentassets/8d19e716636a47c184e7221c77563c93/technology20roadmap20to20205020no20foreword.pdf

IATA. (2019b). *Fact sheet fuel.* Retrieved March 9, 2020, from https://www.iata.org/contentassets/e946531e45da4a1a928f01a908a4a3aa/fact-sheet-fuel.pdf

IATA. (2020a). *Current trends. 2019-2039.* Retrieved June 8, 2020, from https://www.iata.org/contentassets/e938e150c0f547449c1093239597cc18/pax-forecast-infographic-2020-final.pdf

IATA. (2020b). *Developing Sustainable Aviation Fuel (SAF).* Retrieved May 16, 2020, from https://www.iata.org/en/programs/environment/sustainable-aviation-fuels/

IATA. (2020c). *Five years to return to the pre-pandemic level of passenger demand.*

IATA. (2020d). *Managing cabin waste.* Retrieved April 16, 2020, from https://www.iata.org/en/programs/environment/cabin-waste/#tab-1

IATA. (2021). *Climate change.*

ICAO. (2016). *Environmental report 2016. Aviation and Climate Change. On Board a Sustainable Future.*

ICAO. (2019). *Environmental report 2019. Aviation and environment. Destination green: The next chapter.*

IPCC. (1999). *Aviation and the global atmosphere.* Cambridge University Press.

Japan Rail Pass. (2019). *The Japanese Maglev: World's fastest bullet train.* Retrieved June 26, 2020, from https://www.jrailpass.com/blog/maglev-bullet-train

Jittrapirom, P., Caiati, V., Feneri, A.-M., Ebrahimigharehbaghi, S., Alonso, M. J., & González, and Jishnu Narayan. (2017). Mobility as a service: A critical review of definitions, assessments of schemes, and key challenges. *Urban Planning, 2*(2), 13. https://doi.org/10.17645/up.v2i2.931

Johnson, P. (2015). Avoiding decision paralysis in the face of uncertainty. *Harvard Business Review.*

Kettunen, T., Hustache, J.-C., Fuller, I., Howell, D., Bonn, J., & Knorr, D. (2005). Flight efficiency studies in Europe and the United States. In *Proceedings of the USA/FAA Air Traffic Management R&D Seminar 2005, FAA/EUROCONTROL.* Baltimore, MA.

Kirig, A. (n.d.). *Tourismus Nach Corona: Alles Auf Resonanz!* Retrieved July 5, 2020, from https://www.zukunftsinstitut.de/artikel/tourismus-nach-corona-alles-auf-resonanz/

Klimaschutzportal. (2020). *Kerosinverbrauch Senken - Energieeffizienz Steigern Am Flugzeug.* Retrieved February 2, 2020, from https://www.klimaschutz-portal.aero/verbrauch-senken/am-flugzeug/

Linden, E. (2021). Pandemics and environmental shocks: What aviation managers should learn from COVID-19 for long-term planning. *Journal of Air Transport Management, 90*, 101944. https://doi.org/10.1016/j.jairtraman.2020.101944

Lufthansa Group. (2019). *BALANCE. Sustainability Report 2019. Environmental Report*, p 25.

Lufthansa Group. (2020a). *Four pillars for climate protection.* Retrieved December 9, 2020, from https://www.lufthansagroup.com/en/responsibility/climate-environment/fuel-consumption-and-emissions/four-pillars-for-climate-protection.html

Lufthansa Group. (2020b). *Q2 2020 results analyst and press conference call.*

Mayer, R., Ryley, T., & Gillingwater, D. (2014). The role of green marketing: Insights from three airline case studies. *The Journal of Sustainable Mobility, 1*(2), 46–72. https://doi.org/10.9774/GLEAF.2350.2014.no.00005

Mayring, P. (2020). Qualitative Inhaltsanalyse. In G. Mey & K. Mruck (Eds.), *Handbuch Qualitative Forschung in der Psychologie.* Springer.

Morningstar. (2021). *Sustainable funds U.S. landscape report. More funds, more flows, and impressive returns in 2020.*

Morris, H. (2017, June 8). The fastest passenger plane in the sky? It Might Surprise You. *The Telegraph.*

NATS. (2020). *Aviation Index 2020.* Retrieved May 21, 2021, from https://www.nats.aero/news/aviation-index-2020/

Peter, M., Lückge, H., Killer, M., & Maibach, M. (2016). *Auswirkungen Eines EHS-Linking Für Den Bereich Luftfahrt – Aktualisierung Für Die Schweiz. Schlussbericht.* INFRAS.

Rauch, C. (2020). *Leisure Travel: Tourismus Der Zukunft.* Retrieved May 22, 2020, from https://www.zukunftsinstitut.de/artikel/tourismus/leisure-travel-tourismus-der-zukunft/

Rensch, C. (2019). *SBB Prüft Neue Nachtzugverbindungen. Klimafreundliches Reisen.* Schweizer Radio Und Fernsehen. Retrieved June 15, 2020, from https://www.srf.ch/news/schweiz/klimafreundliches-reisen-sbb-prueft-neue-nachtzugverbindungen

Robinson, J. (1982). Energy backcasting: A proposed method of policy analysis. *Energy Policy, 10*, 337–344.

Samadi, S., Gröne, M.-C., Schneidewind, U., Luhmann, H.-J., Venjakob, J., & Best, B. (2017). Sufficiency in energy scenario studies: Taking the potential benefits of lifestyle changes into account. *Technological Forecasting and Social Change, 124*, 126–134. https://doi.org/10.1016/j.techfore.2016.09.013

Schäpke, N., & Rauschmayer, F. (2014). Going beyond efficiency: Including altruistic motives in behavioral models for sustainability transitions to address sufficiency. *Sustainability: Science, Practice and Policy, 10*(1), 29–44. https://doi.org/10.1080/15487733.2014.11908123

Schmidt, M. (2008). Die Bedeutung Der Effizienz Für Nachhaltigkeit – Chancen Und Grenzen. In S. Hartard, A. Schaffer, & J. Giegrich (Eds.), *Ressourceneffizienz im Kontext der Nachhaltigkeitsdebatte* (pp. 31–46). Nomos.

Setchell, M. (2018). *Strategies for airline sustainability.* EMG. Retrieved January 5, 2021, from https://www.emg-csr.com/airline-sustainability/

Solar Impulse Foundation. (2016). *Our story.* Retrieved March 23, 2020, from https://aroundtheworld.solarimpulse.com/our-story

SRF. (2020). *Fliegen Und Autofahren Werden Teurer - Nationalrat Sagt Ja.*

Statista. (2020). *Anzahl Der Passagiere Ausgewählter Europäischer Fluggesellschaften Im Jahr 2019.* Retrieved June 20, 2020, from https://de.statista.com/statistik/daten/studie/29174/umfrage/anzahl-der-passagiereeuropaeischer-fluggesellschaften/

SWISS. (2020). *Airtrain.* Retrieved May 24, 2020, from https://www.swiss.com/ch/en/book/partner-offers/airtrain

SWISS. (n.d.). *Environmental responsibility*. Retrieved March 24, 2020, from https://www.swiss.com/corporate/en/company/responsibility/environmental-responsibility

Teece, D., Pisano, G., & Shuen, A. (1997). Dynamic capabilities and strategic management. *Strategic Management Journal, 18*(7), 509–533.

Teece, D., Raspin, P. G., & Cox, D. R. (2020). Plotting strategy in a dynamic world. *MIT Sloan Management Review, 2020*(3).

Thaler, R., & Sunstein, C. R. (2008). *Nudge. Improving decisions about health, wealth, and happiness*. Yale University Press.

The Natural Step. (2011). *Applying the ABCD method*. Retrieved May 26, 2021, from https://www.naturalstep.ca/abcd

Transport & Environment. (2017). '*Electrofuels - What role in EU transport decarbonisation?* Retrieved August 1, 2020, from https://www.transportenvironment.org/sites/te/files/publications/2017_11_Briefing_electrofuels_final.pdf

Trocmé, J. (2020). *Aviation: The extreme of the extremes*. Retrieved February 18, 2021, from https://insights.nordea.com/en/business/finnair-ceo-interview/

UIC. (2018). *High speed rail. Fast track to sustainable mobility*. Retrieved June 15, 2020, from https://uic.org/IMG/pdf/uic_high_speed_2018_ph08_web.pdf

UVEK. (2018). *Strategische Ziele Des Bundesrates Für Die SBB AG 2019–2022*. Retrieved June 15, 2020, from https://www.uvek.admin.ch/uvek/de/home/uvek/bundesnahe-betriebe/sbb/strategische-ziele.html

Walls, J. L. (2018). The power of one: Leadership and corporate sustainability. *The European Business Review,* (September–October), 59–62.

Walls, J. L., Phan, P. H., & Berrone, P. (2011). Measuring environmental strategy: Construct development, reliability, and validity. *Business & Society, 50*(1), 71–115. https://doi.org/10.1177/0007650310394427

Wittmer, A., Bieger, T., & Müller, R. (Eds.). (2021). *Aviation systems: Management of the integrated aviation value chain*. Springer.

Zeit. (2020). *Lufthansa Verschärft Sparkurs Bei Flotte Und Personal*.

Controlling, Guiding and Assisting: The Role of Airports in the Transition Towards Environmentally Sustainable Aviation

Ivan Vuckovic and René Puls

Abstract

- Aviation is a system and depends on the successful environmental transition of all system members to become environmentally sustainable.
- When it comes to reducing their own Scope 1 and 2 greenhouse gas emissions, airports have mature remedies at their disposal that have already resulted in notable successes.
- Furthermore, airports are also uniquely positioned to assist the other system members on their path to environmental sustainability using 11 distinct levers to build comprehensive action plans adapted to the airport's own unique context.
- In doing so, airports have a direct and indirect influence on aviation's environmental sustainability and benefit cost savings, reduced damages from natural disasters as well as improved public perception.

1 Introduction

With aviation's growth trend expected to recover following the COVID-19 pandemic (see Sect. 5 in Chapter "Sustainable Aviation: An Introduction"), the rising demand for aviation services will again necessitate evermore creative ways of impact mitigation, better technology and increased cooperation between aviation and

I. Vuckovic (✉)
University of St. Gallen, St. Gallen, Switzerland
e-mail: i.vuckovic@hotmail.com

R. Puls
Center for Aviation Competence, University of St. Gallen, St. Gallen, Switzerland
e-mail: rene.puls@unisg.ch

societal stakeholders. Adding to the complexity, the aviation industry is a well-tuned system where all parts have to perform their role for the system as a whole to successfully function and deliver aviation's socio-economic benefits. After all, an airline is worthless without the supporting infrastructure of airports as well as ground handlers and vice versa. Therefore, it becomes clear that all components of the system have to commit themselves to environmental sustainability for the whole system to achieve a successful transition.

Consequently, in order to maintain their long-term profitability as well as ensure the safety and utilisation of their assets, airports have a strong interest in the industry's transition to environmental sustainability. For example the airports' operations are directly threatened by climate change with valuable infrastructure being at risk of natural disasters. However, striving for environmental sustainability can also bring financial benefits as airports identify more and more opportunities to reduce costs, while becoming more sustainable through the implementation of new technologies and processes. Finally, airports face tremendous public pressure and media scrutiny, with climate protests being organised at airports across Europe with at least one protest resulting in legal action against the expansion of Heathrow Airport. Therefore, sustainability goes far beyond risk mitigation and is also beneficial for business.

As a result, environmental protection is a challenge that must be taken seriously by airports, which is why it is not surprising to see a strong environmental mobilisation among airports worldwide, which goes beyond the targets set by the broader aviation industry of halving net aviation carbon emissions until 2050 compared to 2005 levels (Air Transport Action Group, 2018, p. 7). For instance, members of Airports Council International (ACI) Europe have committed themselves to the goal of having 100 carbon neutral airports based on operations fully within their control in 2030, with Swedavia's airports aiming to reach this milestone by 2020 already (ACI Europe, 2017, p. 1; Wennberg, 2019, p. 168).

However, this only reflects the impact directly caused *by* the airport, which disregards significant sources of pollution *at* airport premises. While the airport's own operations do generate emissions, waste and other negative environmental impacts, this is vastly superseded by the environmental footprint of other aviation stakeholders at the airport such as airlines, passengers and commercial partners (e.g. ground handlers or retailers) that are outside of the airport's direct control and whose impact will be discussed in subsequent sections. In order to address the entire environmental footprint created at an airport, these stakeholders hence also need to reduce their own impact. Therefore, an airport has an interest in other actors reducing their footprint, as they would also benefit from these actions.

Airports are thus in a paradoxical position. On the one hand, the majority of the impact caused at the airport premises is not caused by the airport itself but by other industry stakeholders. On the other hand, since the airport's long-term economic success depends on the industry's transition to environmental sustainability, it relies on those same stakeholders to successfully transform their operations to become more sustainable. Consequently, the airport's perspective on aviation's environmental sustainability goes beyond reducing its own direct footprint and includes assisting other aviation stakeholders on their path to a greener future.

Fortunately, as a central piece of the aviation system where almost all major stakeholders meet, airports are uniquely positioned to engage with other stakeholders to improve their own environmental footprint, hence playing a more significant role in aviation's transition to environmental sustainability than just reducing the impact of their own activities. As a result, this chapter presents levers that an airport can utilise to influence the behaviour of airlines, passengers and commercial partners to reduce their environmental impact. In order to preserve and emphasise the socio-economic benefits of airports, it is of vital importance to not only address the negative environmental footprint generated by airports, but also by other impact-producing stakeholders at the airports.

As the first chapter delineated, the COVID-19 pandemic has had a dramatic effect on aviation globally. While demand for aviation has plummeted during 2020 and growth is expected to be dampened in the short to medium term, both are expected to recover in the long term. Therefore, sustainability will remain a vital topic for airports in the future, which is why preparation and a sustainability strategy are still of paramount importance. However, the concrete measures applied by airports will change depending on their financial performance.

2 Environmental Sustainability

The most significant sources of the airport's environmental footprint include emissions, noise, water, biodiversity as well as waste and can be divided into local and global impacts based on the reach of their consequences (Budd et al., 2020, pp. 56–57; Urfer & Weinert, 2011, p. 128). However, before delving deeper into each dimension, it is important to distinguish between the scopes of environmental impact generated at an airport. ACI's (2009, p. 15) adaptation of GHG Protocol's framework does so by differentiating between sources owned or under the control of the airport (Scope 1 and 2) and those that are not (Scope 3). While this segmentation was originally developed for greenhouse gas (GHG) emissions, the notion of separating impact according to sources owned or controlled by the airport operator and sources that are not is intuitive and appealing for other environmental dimensions such as noise, water, biodiversity and waste (Fig. 1).

Impact-Producing Stakeholders
The first chapter introduced the Aviation System and its stakeholders to illustrate the interdependencies that govern the industry. This multidimensional nature of aviation necessitates an integrative approach to ensuring the industry's environmental sustainability that includes all stakeholders from both the supply and demand systems. However, while all stakeholders have a part to play in aviation's transition, only some produce environmental impacts such as emissions or waste at the airport location.

For instance, the regulator is an essential stakeholder in the system, but does not produce noise, emissions, waste or any other impact at the airport premises. The same holds true for various industry associates. Consequently, the impact-producing

Scope 1	Scope 2	Scope 3
Emissions from sources owned or controlled by airport.	*Emissions from off-site electricity generation purchased by the airport operator.*	*Emissions from airport-linked activities from sources not owned or controlled by airport.*
Example: - Fleet vehicles - Airport maintenance - On-site waste disposal - Power plant	Example: - Off-site generation of electricity	Example: - Airplane main engines - APUs - Ground access - Ground support equipment

Fig. 1 Emission sources per Scope. Adapted from ACI (2009, pp. 15–16)

stakeholders that this chapter refers to are those stakeholders whose activities contribute to the airports' Scope 3 impact. Research presented in the subsequent sections typically highlights airlines, commercial partners and people, including both passengers and employees, as the main impact-producing stakeholders, with a footprint often far beyond that of airports.

Emissions

Emissions are the most prominent sources of environmental impact at airports and aviation in general, whose effects can be felt both locally and globally depending on the emitted gas. The airport itself produces emissions primarily through energy consumption, vehicles/ground equipment belonging to the airport, as well as waste and water management (Airports Council International, 2019, p. 11). Due to their size, architectural layout and operational requirements, airports need vast amounts of energy to operate safely and comfortably for up to 24 h/day, hence producing substantial CO_2 emissions as a byproduct (Ortega Alba & Manana, 2016, p. 361).

However, the airport's direct emissions are vastly overshadowed by Scope Three emissions caused by impact-producing stakeholders, such as the aircrafts' landing and take-off (LTO) movements as well as ground access and on-site vehicle activity (Ashford et al., 2013, pp. 534–535; Gatwick Airport, 2019, p. 3; Postorino & Mantecchini, 2014, pp. 84–85). Consequently, Scope One and Two emissions accounted for only 12.03% of total GHG emissions at Changi airport in 2017/2018, while Scope Three emissions were responsible for the remaining 87.97% (Changi Airport Group, 2018, pp. 84–86). When combined with the fact that airports cause roughly 2–5% of aviation's global emissions, it becomes evident that Scope One and Two emissions are only minor contributors to aviation's overall footprint (ACI Europe, 2019a, p. 13) (Fig. 2).

Noise

Noise is a local issue of exceptional importance for airports as it is typically the likeliest environmental dimension to cause strong reactions from local communities, hence making it very controversial and politically sensitive (Budd, 2017, p. 287). In fact, noise levels are a central point of argument when it comes to airport expansions,

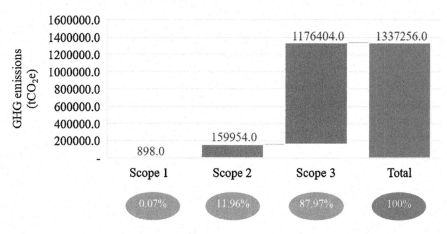

Fig. 2 Singapore Changi GHG emissions scope breakdown in 2017/2018. Adapted from Changi Airport Group (2018, pp. 84–85)

which is why it is often seen as a constraint for operations, capacity and growth, thus making any reductions extremely beneficial to airport operations (Suau-Sanchez et al., 2011, p. 283). The negative impact of noise is so large that one study estimated it to be USD 23.8 billion annually from 181 airports in 2005 based on the willingness to pay for noise abatement (He et al., 2014, p. 98).

Once again, the airport itself is a rather minor noise source compared to the impact of ground access, feeder traffic as well as aircraft. Nevertheless, the airport bears a significant amount of the economic costs when dealing with the consequences by organising noise abatement programmes or foregoing future expansion plans. Therefore, airports would benefit immensely from measures targeted at reducing noise levels at the source by affecting the behaviour of impact-producing stakeholders.

Water

In a sample of nine large international airports, such as Frankfurt or San Francisco, the average water consumption amounted to almost 1 billion litres of water per year (Kilkiş & Kilkiş, 2016, p. 254). This water is used by airports in their operations for drinking, catering and retail operations, cleaning, maintenance and other purposes (Budd, 2017, p. 292). Not only is this a large economic burden for the airport, but it is also an environmental one for the regional water system, thus making water conservation and management highly important.

Biodiversity

While airport emissions also have an indirect effect on biodiversity through climate change, the main impact considered here is the product of the airports' direct operations, such as construction. Airports tend to be located outside of the main urbanised area they are serving and as such are surrounded by rich habitats making the impact highly local. Moreover, to ensure smooth and safe operations, many airports have to follow rules that are detrimental to the local plant and animal species, such as local bird populations (Ashford et al., 2013, p. 544).

Waste

Airports and their impact-producing stakeholders generate substantial amounts of waste from sources such as retailing, construction, flights and maintenance (Budd, 2017, p. 292; Budd et al., 2020, p. 57). For a sample of nine large airports, waste production ranged from 4300 tons (Istanbul) to 26,700 tons (San Francisco) per year with a recycling rate between 13% (Istanbul) and 83% (Frankfurt) (Kilkiş & Kilkiş, 2016, p. 255). As the landlord and central component of the aviation system, airports are expected to run waste management programmes, including the underlying strategy, as well as monitoring and reporting. Consequently, similar to other dimensions, waste management at airports necessitates engagement with the above-mentioned stakeholders in order to be successful.

In conclusion, environmental sustainability at the airport is incredibly complex and includes several stakeholders, in addition to the airport itself, who produce a negative footprint across the five dimensions. While the airport's footprint is considerable and needs to be tackled, there is a larger amount of impact stemming from Scope Three sources, especially when it comes to emissions and noise, that is outside of the airport's direct control. Therefore, airports must take a dual approach to eliminating the environmental impact produced at the airport premises. Where airports control the source of the impact, they must implement *direct measures* swiftly and with determination to mitigate it. However, where there is an impact-producing stakeholder behind the source of the impact, airports need to come up with more *indirect measures* geared towards guiding the behaviour of stakeholders to become more environmentally sustainable.

3 Strategies to Reduce Airports' Direct Environmental Impact

The airport's direct influence on aviation's environmental sustainability revolves around mitigating its own Scope One and Two impact across the five environmental dimensions. In spite of being greatly exceeded by Scope Three impacts, it is still of vital importance to deal with them in order to holistically improve aviation's footprint. In addition, eliminating this impact often carries direct financial benefits for airports as they simultaneously cut costs through lower resource consumption or maintenance fees.

Airports share some sustainability challenges such as temperature regulation, lighting and construction waste with other large pieces of infrastructure, making them comparable to stadiums, train stations, megamalls, large office buildings, harbours, etc. Consequently, the technologies, processes and skills required to make airports environmentally sustainable are often developed and utilised for other use cases that can be translated to an airport context. This results in the availability of mature solutions such as technologies and processes that allow airports to already significantly reduce their footprint today. Of course, there are added complexities posed by safety and security requirements or 24/7 operations, but airports across the globe are still making great strides towards minimising Scope One and Two impacts.

Emissions

Since the majority of direct emissions are caused by the consumption of electricity and fuel, airports rely on energy-saving remedies such as improved insulation, replacement of outdated heating or ventilation systems, LED lighting or building management systems. When it comes to energy sourcing, airports have invested in solar farms and wind turbines on the premises as well as chosen to source sustainable electricity from their providers. Another popular remedy is the deployment of low-emission vehicles in the airport's own operations.

Furthermore, implementing a stringent energy management system is one of the most effective ways of curbing emissions stemming from energy consumption, even in developed nations with modern infrastructure (Akyuz et al., 2019, p. 11; Uysal & Sogut, 2017, p. 1396). As a result, several frameworks are in place to identify and address the most significant sources of carbon emissions at airports, such as the ACI Airport Carbon Accreditation Programme, whose goal is to "encourage and enable airports to implement best practices in carbon management" (Airports Council International, 2019, p. 8). Consequently, it establishes a process that helps airports identify carbon-emitting operations, map their effect, devise action plans for their management, and then execute those plans.

In doing so, airports can achieve four levels of accreditation based on whether they (1) map their carbon footprint, (2) actively plan and execute reduction activities for Scope One and Two emissions, (3) optimise stakeholder engagement to reduce some Scope Three emissions as well as (4) offset all carbon emissions that cannot be reduced. In 2019, ACI has accredited a total of 274 airports worldwide, covering 43% of global passenger flows and with 50 airports reaching carbon neutrality (Airports Council International, 2019, p. 6). The first results are already on the horizon since Swedavia, the Swedish state-owned airport operator has committed itself to completely eliminating carbon emissions of their own operations by 2020 (Wennberg, 2019, p. 167). Another 100 airports are following suit until 2030, which is 20 years before the 2050 reduction commitment of the aviation industry (ACI Europe, 2017, p. 1) (Fig. 3).

Noise

Noise as a term describes an *unwanted* and *intrusive* sound meaning that its perception is highly subjective and difficult to quantify, unlike sound, which can be objectively quantified (Budd, 2017, p. 288). This subjectivity may also result in a low correlation between recorded sound levels and perceived annoyance (Budd, 2017, p. 289). Therefore, while the impact of noise definitely depends on the physical level of sound, non-acoustic elements such as the subjective perception of noise and subsequent annoyance play a significant role, too (ACI Europe, 2019b,

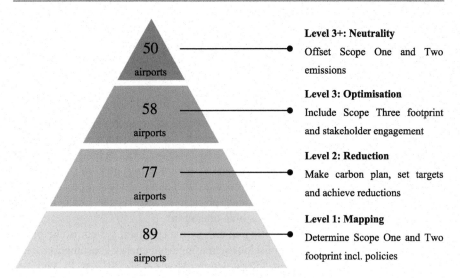

Fig. 3 ACI Carbon Accreditation levels. Adapted from Airports Council International (2019, pp. 9, 20)

p. 29). In fact, policies aiming to reduce noise annoyance rather than the actual sound levels may actually be more effective in combatting noise pollution.

As a result, there are two principal ways of dealing with the noise question at an airport: either the sound level or the annoyance is reduced. Looking at the sound-based measures, they are almost all indirect in nature because the airport is not the one producing the actual sound and as such will be covered in the subsequent section. Nevertheless, airports can, for instance, urge and help regulators zone land in a way where no future residences will be built in areas impacted by noise by advising them and offering reliable metrics (CANSO & Airports Council International, 2015, pp. 22–23).

When considering annoyance-based approaches, research suggests that providing transparent and sufficient information is helpful as it allows people to plan for the noise in their decision-making and ensures predictability (Phun et al., 2016, p. 1). Therefore, engaging and informing local communities on noise-related issues is critical for an effective environmental sustainability strategy in order to correctly meet their needs (ACI Europe, 2019b, p. 29). Noise insulation schemes offer another approach for reducing the annoyance of the affected residents (ACI Europe, 2019a, p. 14). Furthermore, while a wider region enjoys the benefits of airports, the noise-related costs are largely absorbed by a much smaller local community, hence creating a disbalance. As a result, airports should distribute the benefits towards local residents through employment or investment opportunities (Ashford et al., 2013, p. 531).

Water

By minimising water use at the source with collection systems, automatic regulation, waterless sweepers and other actions, airports are able to achieve large reductions in water consumption, reaping economic benefits in turn (Ashford et al., 2013, p. 537). Measures such as dual drainage systems allow airports to re-use greywater from non-industrial purposes, such as laundry, for other purposes like toilet flushing or landscaping (Ashford et al., 2013, p. 537). Research shows that, due to the large demand for non-potable water, airports are promising environments when it comes to the recycling of greywater with simple and inexpensive systems already being available (Couto et al., 2015, pp. 377–378). Moreover, the airports' demand for non-potable water can also be supplied by rainwater in areas with sufficient rainfall (Moreira Neto et al., 2012, p. 42).

The monitoring and management of surface drainage are of paramount importance for airports so as to mitigate pollution caused by fuel, oils, de-icing chemicals or any other hazardous material (Budd, 2017, p. 291). After all, if airports are not careful, run-offs of warm water or glycol can contaminate nearby water systems and culminate in dead zones for aquatic life or even threaten human health (Ashford et al., 2013, p. 540). Fortunately, surface and groundwater pollution can and must be properly managed through appropriate infrastructure and operational practices, usage of environmentally friendly chemicals as well as better planning and handling procedures (Ashford et al., 2013, p. 541).

Biodiversity

Due to the criticality of wildlife management for their operational safety, airports have historically paid high attention to it, especially in the case of birds (Altuntas, 2019, p. 89; Hesse et al., 2010, p. 185). Therefore, biodiversity management strategies are well-established in practice and include a series of policy options that are frequently stipulated and defined by local regulations, such as sound systems for scaring birds, dedicated nature conservation zones and relocation programmes, among others (ACI Europe, 2019b; Hesse et al., 2010, pp. 185–186). Airports are going even further to preserve and stimulate the local biodiversity by recording on-site species, setting aside land for conservation and engaging the local community in ongoing projects (Bicker, 2018, p. 4). A sample of some of the busiest and best airports in the world showed that, on average, airports keep 561 hectares for the conservation of biodiversity (Kilkiş & Kilkiş, 2016, p. 255).

Waste

Similar to other industries, the standard approach of reducing, reusing and recycling waste applies to airports as well. However, in contrast to many other industries, airports also need to deal with waste coming from all over the world as well as harmful materials such as chemicals which require special care, hence increasing complexity (Ashford et al., 2013, p. 538). Nevertheless, going beyond the recycling of passengers' trash, airports have found other innovative ways of recycling materials consumed at the airport. For instance, Phoenix Sky Harbour International Airport recycles runway friction rubber, concrete from runways and taxiways, as well as building materials from demolished properties, bringing about substantial financial benefits while lowering emissions and material waste (Lissner, 2014, pp. 108–110).

Even though Scope One and Two are often outpaced by Scope Three impacts, airports are well positioned to eliminate large portions of their direct footprint through the many direct measures available to them. They benefit from the ACI, a strong representative body that has created a framework for the identification, monitoring and mitigation of carbon emissions, thus providing a comprehensive guide for airports. Next, environmental leaders have already emerged among airports that others can mimic and learn from, thus seeing what works and what not. Finally, a great share of the technology, processes and skills required to greenify airports are also applicable to other use cases outside of aviation, which sped up their development and resulted in a higher maturity that airports benefit from now. As a result, the tools needed to mitigate Scope One and Two impacts exist, and airports are ready to use them.

4 Strategies to Reduce Airports' Indirect Environmental Impact

While the direct measures focus on eliminating the airport's Scope One and Two impact, the indirect measures tackle the Scope Three footprint. Moreover, they also help curb the footprint of impact-producing stakeholders beyond the airport as an additional benefit. As a consequence, airports help speed up aviation's transition to environmental sustainability by utilising indirect measures.

Therefore, the question arises of how exactly airports can use their resources, knowledge and central position in the aviation system to help other stakeholders become more environmentally sustainable? To answer this question, this chapter proposes the Airport Sustainability Stakeholder Influence Framework (ASSIF), which reflects the three roles airports play in aviation's greenification and delineates 11 distinct levers that they can use to achieve this goal. First, they can act as controllers ensuring that there are rules and incentives in place that stimulate the transition. Secondly, they can guide stakeholders to environmental sustainability by educating them or signalling the desirable course of action. Thirdly, airports can assist them directly with strategic or operational support (Fig. 4).

The ASSIF is a result of 15 interviews with C-level executives and sustainability managers from both hub and non-hub airports as well as industry experts. All of the airports included in the original study belong to European environmental champions and show impressive strides towards environmental sustainability, hence making them particularly interesting study subjects. Furthermore, later comparisons and triangulation of the interview insights with official sources such as annual or sustainability reports as well as press releases or interviews allowed for data verification, increased accuracy and bias reduction. Coding the interviews and primary documents yielded hundreds of first-level data points that created the basis for the

Fig. 4 Airport Sustainability Stakeholder Influence Framework (ASSIF). Own illustration

Table 1 Overview of the sampled airports

Airport	IATA code	Country
Amsterdam	AMS	Netherlands
Basel	EAP	Switzerland
Budapest	BUD	Hungary
Brussels	BRU	Belgium
Copenhagen	CPH	Denmark
Düsseldorf	DUS	Germany
Eindhoven	EIN	Netherlands
Geneva	GVA	Switzerland
Munich	MUN	Germany
Swedavia	Several	Sweden
Zurich	ZRH	Switzerland

subsequent analysis. The categorisation and aggregation of these first-level codes under emerging theories revealed the three major airport roles that build the ASSIF.

All of these levers are already being implemented at airports across Europe, meaning that they are mature and actionable. Consequently, the ASSIF was developed specifically for airport managers to target Scope Three impacts, thus making it a tailored mitigation tool. However, general learnings can also be applied outside of the airport context (Table 1).

4.1 Controlling

The controlling function of airports places the stakeholder's behaviours into focus and tries to directly exert influence on them.

Regulating

Geneva Airport mandates retailers operating at the airport to source a certain percentage of their offering locally.

Regulating as a lever reflects the power of an airport to define the framework of actions of different actors at its premises, i.e. setting the rules of the game. Airports regulate numerous activities for various reasons, such as safety, which is why it is not surprising that rules are utilised for environmental purposes as well. Typically, an airport has two choices when setting a rule—it can either mandate a certain behaviour or ban it.

For instance, airports might require airlines to use only one engine for taxiing or to adhere to continuous descent and glide operations. Another commonly observed example was the introduction of a night curfew or quota system, prohibiting or severely limiting departures and landings during the night. When it comes to commercial partners, airports commonly leverage tendering and procurement procedures or rental agreements to mandate environmentally friendly behaviour. Going even further, it is also possible to prohibit certain activities such as the operation of combustion engine baggage vehicles.

Sometimes the source of regulations is not the airport itself but rather a regulating body or ATC. In this case, airports can lobby these institutions for the desired regulations. However, simply setting rules unilaterally without considering their impact, such as financial consequences, is dangerous. This is especially true for ground handlers who operate under extremely tight margins, hence making top-down implemented measures particularly sensitive.

(Dis)Incentivizing

Kiss and drop fees are used to charge people for pick-ups and drop-offs, hence decreasing their attractiveness as a mode of ground access.

(Dis)Incentivising describes the activity of creating extrinsic (de)motivators for certain behaviours. While implementable for all stakeholder categories, (dis)-incentives are a critical tool for guiding the behaviour of airlines. They primarily utilise financial motivators, hence making them a very appealing and powerful tool in a low-margin industry such as aviation, where incentives can provide a welcome financial injection.

Differentiated charges are one of the most commonly used levers, with numerous airports relying on them to influence airlines' noise and/or GHG emissions, as well as motivating them to land and take-off at certain times of day. In addition to differentiated environmental charges, some airports also offer a bonus or kickback on the passenger tax if airlines offset their emissions. However, skepticism exists regarding the effectiveness of differentiated charges; citing their lack of scale and scope as a drawback. Critics argue that such charges often do not represent a material share of overall charges levied by an airport and would require a centralised approach across all airports to truly stimulate fleet renewals.

Since public transportation is outside of the airports' influence, it is difficult to implement incentives targeted at increasing its share in the airport's modal access split, which is why airports typically rely on disincentivising the other modes of transportation. Consequently, in addition to the introductory example, airports can also raise parking charges in order to stimulate the use of public transportation and thus help decrease the passengers' footprint.

The primary use for (dis)incentives for commercial partners lies in waste reduction. For instance, airports can introduce the "polluter pays principle" where commercial partners are charged for the waste that they produce, thus incentivising recycling and reduced resource consumption. Similarly, several airports increase the fees for the disposal of general waste while taking away recycled waste for free. In doing so, airports can successfully stimulate recycling among commercial partners and thus increase the overall recycling rate at the airport.

Choice Architecture

By introducing a compensation kiosk after the security screenings, an airport nudges passengers to offset their flight while they wait for boarding.

Choice architecture refers to designing potential choices of actions to guide stakeholders' decision-making towards a preferred choice. In other words, choice architecture is used to shape the behaviour of impact-producing stakeholders by shaping their perceived frame of possible actions itself. For example nudging is a frequently used method in choice architecture used to establish a predictable pattern of behaviour without outright banning certain options.

As sustainable aviation fuel (SAF) becomes an increasingly promising innovation for reducing aviation's carbon emissions, it also becomes more important to create the demand for it. One method airports can implement to support this is by centrally mixing SAF into standard fuel that is offered to airlines. As a result, airlines automatically receive a fuel blend as airports make this choice for them.

Choice architecture has proven to be especially effective in influencing the behaviour of people. Looking at nudges, airports offer prominent dedicated parking spaces for carsharing or electric vehicles to increase their attractiveness. Airports also frequently make the choice for passengers, employees and visitors. For instance, by only allowing electric taxis to operate at the airport, Schiphol limits the choice of people arriving by taxi in turn lowering their footprint.

Infrastructural improvements such as dedicated charging stations can provide a needed nudge for ground handlers to switch to electric vehicles as they can be sure that their fuel needs will be met. Furthermore, energy management is another field where this lever can be applied effectively. By centrally determining the energy mix and opting to purchase more renewable energy, an airport makes a choice on behalf of all of its commercial partners that ultimately lowers their environmental footprint.

This lever presents an exciting field of influencing as airports look into novel ways of guiding behaviour beyond regulations and (dis)incentives. While nudging has dominated this lever so far, other options such as green defaults exist and can be implemented by airports as the application of this lever matures. Finally, choice architecture is a promising alternative to regulations in many cases as it does not require control mechanisms nor punishments to work.

Monitoring

 Airports can require a progress report from partners to ensure that they measure and can prove the impact of their actions.

Monitoring as a lever refers to the influence function of tracking and controlling stakeholder behaviours. While its primary purpose is often data acquisition as well as ensuring that regulations are being followed, monitoring also has an independent guidance function. After all, simply knowing that one's behaviour is being tracked is enough to act in an "acceptable" way without any underlying rules commanding it.

Although all airports engage in some sort of environmental impact monitoring, most commonly emissions and noise, not all of them track these dimensions for their impact-producing stakeholders. Both noise and emissions are typically measured and calculated. While measuring tells the real story of local pollution, calculations help determine the global impact. Furthermore, measurements are most often not attributable to individual actors since a sensor cannot determine whether a GHG came from an aircraft, a ground vehicle or nearby motorways, which is why calculations can help fill that gap.

Both the noise and emissions produced by aircraft are publicly available at numerous airports, thus putting the impact of individual airlines on display, which then motivates them to improve their behaviour. London Heathrow has even created a ranking of airlines operating at the airport according to noise and emissions KPIs, which is also available to the public (Heathrow Airport, 2020). Similar to those of airlines, emissions for ground vehicles are also typically calculated.

Currently, the main challenge facing the effectiveness of monitoring as an influencing lever lies in technology since it is not fully advanced to comprehensively cover taxiways, remote stands and the apron. Upcoming communication technology such as 5G will solve this, hence ushering in improved monitoring capabilities and, ultimately, more transparency. This will then allow airports to measure actual emissions and noise instead of calculating them, which is already being considered at some airports.

4.2 Guiding

While the controlling function places the focus on the impact-producing stakeholders, the guiding function considers the behaviour of the airport more closely. It looks at behaviours airports can adopt to influence the environmental conduct of impact-producing stakeholders.

Communicating

> Zurich Airport communicated the maturity of sustainable aviation fuel technology by sponsoring the fuel used for the 2020 World Economic Forum.

The Communicating lever uses discussions and interactions with stakeholders to guide them towards desired behaviours. In doing so, it primarily deals with *why* questions to generate buy-in, motivation and need for action. This in turn helps impact-producing stakeholders rally behind the same cause and ensures the spread of critical information. Communication generates an internal source of reasoning and behavioural change among stakeholders, which is not only a very powerful tool in itself but also a notable complement to external sources such as regulations or (dis)incentives. Moreover, engaging in open and transparent communication generates trust between parties, hence making it easier to adapt one's behaviour.

The most obvious example of communication can be found in the consultation process between airports and airlines regarding airport charges, where airports need to explain why there is a charge, what it will be used for and what the benefits of it will be. Communication allows airports to create buy-in, thus generating voluntary behavioural changes from airlines. For instance, Geneva Airport and its airlines successfully avoided unilateral actions from the regulator by pre-emptively implementing night curfews on joint terms.

With passengers, airports typically communicate the benefits, or the why, of public transportation or other initiatives such as shopping on arrival. By explaining the benefits both for the traveller and the environment, passengers are more open to changing their behaviour to the one preferred by the airport. Additionally, communicating general airport initiatives and why they are important to both passengers and employees ensures that people are aware of them and utilise them, as in the case of self-service check-in or baggage drop.

Educating

> By educating ground handlers on administrative tasks regarding governmental subventions for e-vehicles, Budapest Airport can help them electrify their fleet.

Educating relies on teaching information or skills that help impact-producing stakeholders lower their footprint. While communicating deals with the *why* question, educating tackles the *how*, which is why these two measures complement each other.

When it comes to educating airlines, airports can organise studies or pilot projects to help explore new concepts. For example performing feasibility studies on electric taxiing and setting up pilot projects for some initiatives helps airlines gain data and experience they need to turn ideas into reality. Airports are also well-positioned to educate passengers on actions to reduce their general travel footprint, such as reducing the weight of their luggage.

Budapest Airport engaged in very creative educational techniques for increasing the share of electric vehicles at the airport. First, the airport provided firsthand experiences on electric vehicles to ground handlers, since it had acquired some for itself beforehand. Second, it also educated ground handlers on how to handle the administrative tasks needed to receive governmental subventions. Third, to convince taxi operators to switch to electric vehicles, the airport extensively studied their business model to develop a financial model underscoring the benefits of a transition. While there is no economic incentive for an airport to do this, it benefits from the increased share of electric vehicles at the airport and, in turn, lower emissions.

As demonstrated in the examples outlined above, education offers guidance to stakeholders and lowers the hurdle of changing behaviours, thus motivating them to transition to more environmentally sustainable behaviours. Of course, a stakeholder such as an airline has much deeper insights into their own challenges, operations and possibilities, but having a capable, external view is also very beneficial. After all, airports are an integral part of many of the operations run by stakeholders, so they also have the required expertise. Moreover, a different viewpoint is critical for generating innovative ideas and critically examining existing practices.

Signalling

 Introducing environmental KPIs that are as important as financial ones, signals to the airport's stakeholders that environmental sustainability is a priority.

Signalling refers to airports credibly illustrating their intents or any other information on environmental sustainability to others. The purpose of this lever is to motivate stakeholders to change their behaviour through reduced uncertainty, increased transparency and demonstrated commitment to lowering the airport's footprint and the steps needed to accomplish this. As a consequence, stakeholders can calibrate their behaviours to match the airport's signals.

One of the most prominent airport signalling methods revolves around joining an organisation or alliance dedicated to an environmental goal. The ACI's Carbon Accreditation programme is the best example of this as many airports partake in it in order to showcase their commitment to carbon reduction. Nevertheless, airports can also act alone when it comes to signalling.

This is most credibly achieved by "tying their hands," i.e. making environmental sustainability a priority. One way of implementing this is by making sustainability a core part of their strategy, while sustainability master plans present another way of conveying future initiatives to impact-producing stakeholders. In addition, airports can elevate environmental KPIs for its management to the same level of importance as financial KPIs to illustrate their prioritisation.

With signalling, airports display commitment to environmental sustainability, thus increasing their credibility in negotiations and indicating to other stakeholders to follow suit if they want to stay competitive at the airport premises. As a result, it is useful for aligning stakeholders on a common goal and reducing action paralysis caused by a lack of information. However, signalling depends on trust and credibility, which can be developed using the next lever.

Leading by Example

 Copenhagen Airport invests heavily in making its own infrastructure greener through, e.g., better heating systems or a fleet of electric vehicles.

Leading by example reflects the airport's ability to act as a role model for the impact-producing stakeholders. In other words, it shows that the airport is only asking the stakeholders to do something that it already does itself. In doing so, the airport illustrates firsthand that it is possible and even beneficial to become more environmentally sustainable. There are two main levers that can be utilised by airports: investing in infrastructure and changing own behaviours.

First, an airport can upgrade its heating and cooling systems, thus significantly improving its footprint. Moreover, an airport can also electrify its own fleet of vehicles if it requires the same of its ground handlers. Investments in green energy production facilities also represent a notable way of exercising leadership as well as help commercial partners reduce their own impact by providing them with sustainably sourced energy. The possibilities for infrastructural investments geared towards increased environmental sustainability are numerous and a major focus of existing scientific literature. In summary, such infrastructural activities typically entail significant investments, hence highlighting the airport's commitment to the environmental sustainability of its own operations.

Second, by changing its own behaviours, an airport can also find a way to lead by example. One case can be found in offsetting. By offsetting the emissions of its own operations or those of its employees, the airport motivates its impact-producing stakeholders to do the same. Adopting rigorous environmental management standards, as well as measuring and tracking the own impact, is a further example of leadership that stakeholders can look up to.

In addition to enhanced management practices, the extensive construction activities at the airport premises provide another opportunity for environmental example setting. Ensuring that the materials used are sustainable and properly recycling them is just the tip of the iceberg. Finally, behavioural leadership also includes reaching goals and milestones ahead of the general industry to prove that even more ambitious strategies than the existing ones are possible, as seen in the previously discussed example of Swedavia.

4.3 Assisting

Lastly, levers in this focus area consider measures where the airport provides both material and non-material resources to the impact-producing stakeholder as support.

Cooperating

 Conducting joint audits allows the airport to identify potentials for joint action with stakeholders.

The next lever, cooperating, seizes the opportunity to guide partners towards certain behaviours through the influence on the underlying relationship. In other words, airports can make sure environmental sustainability is included in the common collaboration agenda. As a central part of the aviation system where all stakeholders meet, an airport often has a facilitative function and is even part of projects that it might not be directly affected by. Because of this position, it can participate, initiate or mediate cooperations as visible in the examples outlined below.

When it comes to operations, airports and airlines frequently cooperate as in the case of optimising departure and arrival routes. One of the major enablers of this lies in CEM and A-CDM, which allow the airports to not only significantly improve operational efficiency but also identify promising trial opportunities for more environmentally friendly processes. For instance, the estimated reduction of taxi times due to A-CDM saved more than 100 thousand tons of CO_2 at 17 CDM airports through reduced fuel consumption (Eurocontrol, 2016, p. 23). As a result, A-CDM is capable of significantly reducing the environmental impact at an airport by including the whole ecosystem in one concept and as such can serve as guidance for other multi-stakeholder environmental initiatives.

A-CDM is only one example of a successful collaboration between airport stakeholders. A collaborative approach is the preferred lever when working on sensitive issues such as night curfews and airport charges as it eases buy-in from airlines. Furthermore, airports are successful as facilitators between airlines and energy producers in the production of SAF since they can push all parties in the same direction and offer the required infrastructure.

Airports also cooperate with commercial partners. One potential source lies in mediating the pooling of equipment among ground handlers, which lowers overall investment costs and increases utilisation, hence decreasing the financial burden on ground handlers. This example again underlines the importance of the airport as a neutral facilitator. Next, airports often jointly developed electrification plans with their ground handlers in an attempt to properly define and execute a transition.

Together with retailers, airports can initiate targeted waste reduction programmes such as re-fill stations or discounts for customers who bring their own containers.

Going even further, airports should look for ways to cooperate with other airports on national and global projects, thus engaging in coopetition. For instance, by sharing knowledge and best practices, airports save time on the independent development of sustainability remedies. ASSIF is a great example of this as it is the result of studying several leading airports on the methods they employed for reducing Scope Three impacts. However, airports can also extend the collaboration to include the sharing of resources or joint development of new technologies to cement their position as a guiding beacon for aviation's environmental sustainability.

Moreover, since the work of several stakeholders comes together at an airport, it is also the role of the airport to put all of the initiatives together and ensure the proper exchange of information needed for effective cooperation. After all, airports are a collection of complicated operations conducted by numerous stakeholders, which is why cooperation with those stakeholders is the most effective way of achieving significant improvements. In addition, cooperation is seen as the lever that generates the most buy-in, since all stakeholders contribute voluntarily. Finally, cooperation will grow in importance as there is an increasing number of "win-win" situations where environmental sustainability is beneficial for, and sought after, by multiple stakeholders.

Enabling

 By providing passengers with a simple and ubiquitous recycling infrastructure, they are enabled to sort their waste.

Enabling highlights the airport's potential to provide stakeholders with the prerequisites needed to lower their own environmental impact. It consists of measures that allow stakeholders to take actions that they would be unable to do otherwise, or at least would be more challenging. Airports can do so through the provision of infrastructure as well as changes in operations.

Providing the right ground infrastructure as a way of displacing the aircraft's APU is a popular method of lowering the footprint of airlines. Moreover, airport investments in towing infrastructure have also been successful in curbing airline emissions. The optimisation of taxiways is another LTO-related remedy that seeks to lower the emissions caused by an airline. These infrastructural measures are also complemented by operational actions. For instance, increasing the efficiency of aircraft turnover through revised processes, improved communication infrastructure or general modernisation leads to lower resource wastes. When it comes to ground handlers, the greatest enabler lies in the provision of a fast and stable energy infrastructure for electric vehicles. After all, ground handlers cannot electrify their fleets, unless airports provide the infrastructure needed to maintain and operate the new vehicles.

Through the provision of recycling infrastructure, people can be enabled to sort their waste. Going a step further, measures such as water re-fill stations or re-usable cups allow people to reduce waste production even more. In addition to waste, passengers should also be enabled to lower their impact from ground access. As public transportation is outside airports' domain, airports have to engage in negotiations with the public transport provider to adapt its schedules to accommodate for flight schedules and employee needs. However, airports can also offer own bus services to allow both passengers and employees to lower their GHG emissions.

As a provider of infrastructure, enablement is at the core of the airport's purpose. The airport can be seen as a landlord, so it is in its sphere of influence to provide the right tools and settings for its tenants to flourish. While environmental initiatives and leadership should also be expected from and motivated in stakeholders, they are sometimes limited by the infrastructure and processes provided by the airport. Consequently, seeking opportunities to enable stakeholders to become more environmentally sustainable is indispensable.

Supporting

 Airports offer not only data and their know-how for research projects on reducing environmental impact, but also their infrastructure, e.g., for testing.

The final lever, supporting, describes airport initiatives that aim to directly assist stakeholders and their environmental strategies. While enablement deals with providing the prerequisites for stakeholders to lower their footprint, supporting refers to actively giving stakeholders a "push" through (non-)monetary measures such as subsidies.

When it comes to the support of airlines, airports look towards SAF. Swedavia, for instance, offers cost subsidies for SAF to airlines that refuel at its airports (Swedavia Airports, 2020, p. 1). Moreover, airports can find ways to retrieve money from governmental or European funds to subsidise SAF. The airport also has several possibilities for supporting passengers and employees. By offering a free ticket for public transportation upon arrival, the attractiveness of public transportation is increased dramatically for passengers arriving in Geneva. Furthermore, airports can also fund their own bus lines so that passengers and employees can reach the airport during times when public transport is scarce or unavailable.

Monetary support has also proven popular for supporting commercial partners. Geneva Airport co-invests in electric equipment together with ground handlers or even covers a part of their investments to speed up the electrification of their fleets. Alternatively, airports can take ownership over ground equipment, thus taking on the investment pressure instead of its commercial partners, since an airport is in a stronger financial position. In doing so, the equipment is pooled among ground handlers leading to less surplus infrastructure and in turn a higher utilisation rate. As a result, total investment needs are reduced, thus increasing the financial viability of electrification. Furthermore, some airports also support the fleet electrification of its stakeholders by providing electricity charging for free.

The introductory example illustrates that airports can support stakeholders through research and development if they provide access to information, know-how and their infrastructure. In doing so, the airport becomes a critical facilitator of sustainability innovation and offers resources for the development of potentially groundbreaking technologies. Due to the drop in utilisation caused by the COVID-19 pandemic, several airports have opened their runways and terminals for testing new processes, procedures and systems with the aim of optimising operations and stimulating innovation.

Both from this example and the ones outlined above, it is visible that this lever causes airports to forego income, make own investments and give access to non-financial resources. However, the leading airports will not see this as philanthropy. On the contrary, they will see it as a wise business decision meant to accelerate the transition to environmental sustainability as well as costs needed to ensure the long-term viability of aviation and hence their own infrastructure.

In conclusion, the airport needs and benefits from impact-producing stakeholders reducing their environmental impact. To achieve this goal, the airport can leverage the ASSIF, which reflects the three functions airports can play in aviation's transition to environmental sustainability: controlling, guiding and assisting. These functions are further broken down into 11 levers consisting of numerous actionable measures that are already being implemented by environmentally leading airports. Consequently, these levers present proven ways of lowering the footprint of impact-producing stakeholders at the airport, which can be implemented today.

5 Managerial Implications

Airports already have numerous options for tackling their Scope One and Two impacts at their disposal that are covered both in scholarly and practical literature as well as embedded in successful airport projects across the globe. One reason for this lies in the fact that many solutions are mature and readily available since there are many use cases for them outside of the airport context, such as other major infrastructure projects.

When considering the reduction of the airports' Scope Three footprint, airports can utilise the Airport Sustainability Stakeholder Influence Framework. The ASSIF offers airports a comprehensive set of levers to influence the environmental behaviour of impact-producing stakeholders. In other words, this framework represents the landscape of possibilities airports have at their disposal and can be seen as a catalogue of levers to choose and mix from. While most of the environmentally advanced airports do in fact employ most of the levers to a varying extent, the underlying measures differ significantly between each airport.

This makes sense since each airport operates in its own unique, yet convoluted environment shaped by its location, regulations, business model, strategy, and stakeholder landscape, among other factors. After all, for some airports, noise is the biggest concern, while others see emissions as their greatest environmental challenge. Alternatively, some host several ground handlers, while others only one and some even conduct their own ground handling services.

Consequently, airports seeking to implement these levers need to create measures suitable for their own context. This framework represents a comprehensive catalogue of levers at their disposal for a structured approach to developing concrete measures to tackle their own challenges and achieve their desired goals. Therefore, the ASSIF can be used by all airports no matter their level of environmental sustainability. Environmentally advanced airports can utilise it to complement their existing measures through the insights gained from other leaders, while lagging airports can employ the framework to leapfrog to the level of the leaders.

Airports do not have to implement all levers to successfully lower their Scope Three footprint. Furthermore, the framework does not stipulate what specific levers need to be implemented to tackle an environmental challenge. Nevertheless, airports should not rely on a single lever to affect the environmental behaviour of their stakeholders, but rather combine different levers in order to achieve the best results.

For instance, the electrification of the vehicle fleet of a ground handler is a goal set by many airports. To achieve this, an airport could revise the framework and decide that simply imposing a rule mandating that 20% of vehicles must be electric is the best option for this. However, an airport should rather consider a combination of levers in order to make the transition easier and more effective.

Consequently, it could combine the above-mentioned *regulation* to ensure a minimum result, with support through a *subsidy* so as to lower the financial burden on ground handlers. In parallel, the airport could consider electrifying even more than 20% of its own fleet as a sign of *leadership*. To ensure a faster and cheaper transition for both actors, the airport and ground handler could *cooperate* on a joint application for a grant by the state or EU, as well as *collaborate* on the required financial models. However, a different airport might not have the same resources at their disposal, thus necessitating a different approach.

While aviation demand and growth levels are expected to reach pre-pandemic levels in the long term, the financial pressure caused by COVID-19 will influence the attractiveness of individual ASSIF levers. Therefore, levers that require vast financial contributions by airports without a clear ROI, such as enabling or supporting, will become less attractive. In contrast, levers that utilise non-monetary mechanisms to achieve their goals, such as choice architecture, signalling and cooperating, will gain in importance. This will lead to the development of creative new measures as alternatives to financially incentivising certain behaviours. Nevertheless, airports need to consider the fact that financial incentives will gain in appeal for commercial partners and airlines as they cut their own costs in response to the pandemic.

Finally, the application of this framework carries some strategic consequences for airports as well. By implementing the proposed ASSIF, airports grow to become trusted advisors to aviation stakeholders as well as an indispensable engine and coordinator of the industry-wide environmental transition. As a result, they will solidify their central position in the aviation ecosystem even further as sustainability grows in importance, hence boosting their significance for the industry.

Consequently, the underlying transition from landlord and infrastructure provider to a strategic partner at the nexus of aviation stakeholders will allow airports to boost their overall influence on the industry as well.

6 Conclusion

Airports play a crucial role in aviation's transition to environmental sustainability and it is twofold. First, they have to mitigate their own Scope One and Two footprint for which they bear the main responsibility, resulting in a direct influence on aviation's sustainability. Second, they have a much broader yet indirect effect since they have the opportunity to and interest in assisting other aviation system members in reducing their own impact.

Airports have numerous measures at their disposal for targeting their Scope One and Two footprints. These direct measures have largely achieved maturity, and airports across the globe have shown strong mobilisation to adopt them. They tackle emissions, noise, waste, water and biodiversity, which are the five main sources of environmental impact at an airport, leading to impressive results already. The production of renewable energy, modernisation of heating infrastructure, new building practices or water repurposing are all great examples of how airports are working towards a greener future. In many cases, such direct measures even bring financial gains to the airports through increased efficiency or lower maintenance costs.

When it comes to the indirect influence on aviation's transition, environmentally leading airports have realised that they can help other impact-producing stakeholders reach environmental sustainability through controlling, guiding and assisting. The ASSIF summarises these functions with 11 levers that represent a plethora of concrete actions that are already being implemented. As a result, airports can use this framework to structure and develop innovative measures for their own context or adopt existing ones that meet their needs.

A successful transition of all aviation system members is of utmost importance for minimising the negative consequences on the environment, maintaining the long-term health of the industry as well as ensuring future returns for shareholders. For airports specifically, the benefits include the long-term utilisation of their investments, financial gains from improved efficiency, safeguarding assets from natural disasters and alleviating public pressure, hence putting the focus again on the immense socio-economic benefits of aviation.

References

ACI Europe. (2017). *European airports double their pledge: 100 carbon neutral airports by 2030*. Paris. Retrieved from https://www.aci-europe.org/downloads/mediaroom/170 6133Europeanairportindustrydoublesitspledgeto100carbonneutralairportsby2030 PRESSRELEASE.pdf

ACI Europe. (2019a). *Policy briefing*. Brussels. Retrieved from https://www.aci-europe.org/downloads/resources/ACIEUROPEPOLICYBRIEFING.pdf

ACI Europe. (2019b). *Sustainability strategy for airports*. Limassol. Retrieved from https://www.aci-europe.org/downloads/resources/acieuropesustainability strategyforairports.pdf

Air Transport Action Group. (2018). *Aviation benefits beyond borders*. Geneva. Retrieved from https://aviationbenefits.org/media/166344/abbb18_full-report_web.pdf

Airports Council International. (2009). *Guidance manual: Airport greenhouse gas emissions management*. Montreal. Retrieved from https://store.aci.aero/product/guidance-manual-airport-greenhouse-gas-emissions-management/

Airports Council International. (2019). *Airport carbon accreditation: Annual report 2018–2019*. Abu Dhabi. Retrieved from https://www.airportcarbonaccreditation.org/library/annual-reports.html

Akyuz, M. K., Altuntas, O., Sogut, M. Z., & Karakoc, T. H. (2019). Energy management at airports. In T. H. Karakoc, C. O. Colpan, O. Altuntas, & Y. Sohret (Eds.), *Sustainable aviation* (1st ed., pp. 9–36). Springer. https://doi.org/10.1007/978-3-030-14195-0

Altuntas, H. (2019). Biodiversity management. In T. H. Karakoc, C. O. Colpan, O. Altuntas, & Y. Sohret (Eds.), *Sustainable aviation* (1st ed., pp. 81–96). Springer. https://doi.org/10.1007/978-3-030-14195-0

Ashford, N. J., Stanton, H. P. M., Moore, C. A., Coutu, P., & Beasley, J. R. (2013). *Airport operations* (3rd ed.). McGraw Hill. https://doi.org/10.1017/CBO9781107415324.004

Bicker, R. (2018). *Gatwick biodiversity action plan: Five year review 2012–2017*. London. Retrieved from. https://www.gatwickairport.com/globalassets/business%2D%2Dcommunity/new-community%2D%2Dsustainability/sustainability/gatwick-bap-five-year-review-2012-2017.pdf

Budd, T. (2017). Environmental impacts and mitigation. In L. Budd & S. Ison (Eds.), *Air transport management: An international perspective* (1st ed., pp. 283–307). Routledge.

Budd, T., Intini, M., & Volta, N. (2020). Environmentally sustainable air transport: A focus on airline productivity. In T. Walker, A. S. Bergantino, N. Sprung-Much, & L. Loiacono (Eds.), *Sustainable aviation: Greening the flight path* (1st ed., pp. 55–78). Palgrave Macmillan. https://doi.org/10.1007/978-3-030-28661-3

CANSO, & Airports Council International. (2015). *Managing the impacts of aviation noise: A guide for airport operators and air navigation service providers*. Retrieved from https://www.canso.org/sites/default/files/ManagingtheImpactsofAviationNoise_HQ.pdf

Changi Airport Group. (2018). *Sustainability Report 2017–18*. Retrieved from https://www.changiairport.com/content/dam/cacorp/publications/CAGSustainabilityReport.pdf

Couto, E. A., Calijuri, M. L., Assemany, P. P., Santiago, A. F., & Lopes, L. S. (2015). Greywater treatment in airports using anaerobic filter followed by UV disinfection: An efficient and low cost alternative. *Journal of Cleaner Production, 106*, 372–379. https://doi.org/10.1016/j.jclepro.2014.07.065

Eurocontrol. (2016). *A-CDM impact assessment*. Brussels. Retrieved from https://www.eurocontrol.int/sites/default/files/2019-04/a-cdm-impact-assessment-2016.pdf

Gatwick Airport. (2019). *Carbon: 2018 overview*. London. Retrieved from https://www.gatwickairport.com/globalassets/decade-of-change%2D%2D-carbon-2018-overview.pdf

He, Q., Wollersheim, C., Locke, M., & Waitz, I. (2014). Estimation of the global impacts of aviation-related noise using an income-based approach. *Transport Policy, 34*, 85–101. https://doi.org/10.1016/j.tranpol.2014.02.020

Heathrow Airport. (2020). *Fly quiet and green: The league table*. Retrieved April 10, 2020, from https://www.heathrowflyquietandgreen.com/2019-qtr-3rd/

Hesse, G., Rea, R. V., & Booth, A. L. (2010). Wildlife management practices at western Canadian airports. *Journal of Air Transport Management, 16*(4), 185–190. https://doi.org/10.1016/j.jairtraman.2009.11.003

Kilkiş, Ş. Ş., & Kilkiş, Ş. Ş. (2016). Benchmarking airports based on a sustainability ranking index. *Journal of Cleaner Production, 130*, 248–259. https://doi.org/10.1016/j.jclepro.2015.09.031

Lissner, H. (2014). Emphasising sustainability and exceeding goals. *Journal of Airport Management, 8*(2), 105–113.

Moreira Neto, R. F., Carvalho, I. C., Calijuri, M. L., & Santiago, A. F. (2012). Rainwater use in airports: A case study in Brazil. *Resources, Conservation and Recycling, 68*, 36–43. https://doi.org/10.1016/j.resconrec.2012.08.005

Ortega Alba, S., & Manana, M. (2016). Energy research in airports: A review. *Energies, 9*(5), 349–367. https://doi.org/10.3390/en9050349

Phun, V. K., Hirata, T., & Yai, T. (2016). Effects of noise information provision on aircraft noise tolerability: Results from an experimental study. *Journal of Air Transport Management, 52*, 1–10. https://doi.org/10.1016/j.jairtraman.2015.11.005

Postorino, M. N., & Mantecchini, L. (2014). A transport carbon footprint methodology to assess airport carbon emissions. *Journal of Air Transport Management, 37*, 76–86. https://doi.org/10.1016/j.jairtraman.2014.03.001

Suau-Sanchez, P., Pallares-Barbera, M., & Paül, V. (2011). Incorporating annoyance in airport environmental policy: Noise, societal response and community participation. *Journal of Transport Geography, 19*(2), 275–284. https://doi.org/10.1016/j.jtrangeo.2010.02.005

Swedavia Airports. (2020). *Swedavia – Sustainable aviation fuel (SAF) incentive programme 2020*. Retrieved from https://www.swedavia.com/globalassets/flygmarknad/swedavia-saf-incentive-2020.pdf

Urfer, B., & Weinert, R. (2011). Managing airport infrastructure. In A. Wittmer, T. Bieger, & R. Müller (Eds.), *Aviation systems: Management of the integrated aviation value chain* (1st ed., pp. 103–131). Springer. https://doi.org/10.1007/978-3-642-20080-9

Uysal, M. P., & Sogut, M. Z. (2017). An integrated research for architecture-based energy management in sustainable airports. *Energy, 140*, 1387–1397. https://doi.org/10.1016/j.energy.2017.05.199

Wennberg, L. (2019). Our ambition is high: Running the most climate-smart airport in the world. *Journal of Airport Management, 13*(2), 167–173.

The Role of Public Policy

Nadine Zumsteg and Andreas Wittmer

Abstract

- Aviation is, compared to other industries, considerably less regulated when it comes to environmental externalities.
- Policy measures need to be implemented to reduce greenhouse gas emissions from aviation and internalise aviation's external costs. To achieve this, measures to reduce the demand and to reduce the remaining flights' climate impact need to be considered.
- To diminish the demand, a change of thinking and air transport consumption is required. This could, for example, be achieved through a declaration of emissions on tickets, defining how passengers pay CO_2 charges (e.g. in addition to the final ticket price), regulating the marketing communication for airline offers (e.g. no price communication) and environmental education.
- For a reduction of greenhouse gas emissions per flight, synthetic fuels provide a promising solution. A mandatory blending quota, for example, could secure demand for synthetic fuels.

1 Introduction

The growing air traffic and international mobility demand intensified the conflict between global accessibility and aviation's impact on the climate. To minimise aviation's negative impact on the environment, there is an intense debate on policy

N. Zumsteg (✉)
University of St. Gallen, St. Gallen, Switzerland

A. Wittmer
Center for Aviation Competence, University of St. Gallen, St. Gallen, Switzerland
e-mail: andreas.wittmer@unisg.ch

measures that have to be implemented. According to industry experts and scholars, policymaking and regulation seem to be the most vital drivers for or against sustainable aviation today. Therefore, this chapter answers what could and should be done in terms of policymaking to make aviation more sustainable. This chapter's main impact is a policy framework that reduces greenhouse gas (GHG) emissions from aviation by presenting different policy measures that are already implemented or could realistically be implemented within the next 10 years. According to the Intergovernmental Panel on Climate Change (IPCC) Report published in 2018, the timeframe is chosen since net CO_2 emissions should decrease by 45% till 2030, which means that the society together with the aviation industry cannot wait for new technologies that will make aviation more sustainable in the long-term future. To provide the appropriate background information for the policy framework, international policies concerning climate change and present regulation in the field of aviation policy will be introduced first. Some general challenges facing the design and implementation of policies for sustainable aviation will follow.

2 International Policies Concerning Climate Change

United Nations are the most important global player producing global regulatory frameworks to combat climate change. Already in 1992, the United Nations Framework Convention on Climate Change (UNFCCC) was created. Its goal is to stabilise GHG concentrations "at a level that would prevent dangerous anthropogenic (human-induced) interference with the climate system" (United Nations, 1992, p. 9).

Linked to the UNFCCC is the *Kyoto Protocol*, an international agreement set up in 1997. It was the first protocol with binding reduction goals. For the first commitment period (2008–2012), those industrialised countries that ratified the Kyoto Protocol made commitments to reduce their GHG emissions by an average of 5.2% compared to their emissions of 1990. It covers "emissions of the six main greenhouse gases, namely: Carbon dioxide, Methane, Nitrous oxide, Hydrofluorocarbons, Perfluorocarbons and Sulphur hexafluoride" (UNFCCC, n. d.). For the second commitment period (2013–2020), participating parties agreed on reducing emissions by at least 18% compared to 1990 and adding Nitrogen trifluoride to the GHGs covered. While domestic aviation emissions are part of the national targets, the Kyoto Protocol does not set fixed limits or caps for GHG emissions caused by international aviation. Hence, the national accounts do not account for the emissions from fuel used for international aviation.

The *Paris Agreement*, which went into force in autumn 2016, is a legally binding instrument for the post-2020 period which obliges all nations to reduce their GHG emissions. The aim is to limit the average global warming to a maximum increase of 1.5 °C, or at least considerably less than 2 °C, compared to the pre-industrial period. Furthermore, financial flows should be channelled into low-greenhouse-gas development. The states' ability to adapt to the changing climate should be improved. Every 5 years, each country needs to submit and elucidate a Nationally Determined Contribution. These show the efforts a country intends to achieve to reduce national emissions. The same GHGs are covered in the second commitment period of the Kyoto Protocol. The Paris Agreement does not explicitly mention emissions from international aviation. However, to reach the 1.5 °C scenario, all sectors, including aviation, will need to contribute.

3 Aviation Policy

To better understand the political debate on aviation, we will now take a closer look at aviation policy. Aviation, compared to other industries, faces little regulation when it comes to environmental externalities. However, the industry has been highly regulated since the first years of commercial aviation for safety and strategic reasons. Travels by planes took off during the 1950s. While aviation is still highly regulated in terms of safety, the market was deregulated during the 1970s. The main trigger was the Airline Deregulation Act that came into force in the United States in 1978 and removed statuary control on US airlines. Other industrialised countries followed this liberalisation of air transport services, which led to a decline in fares and dynamic price discrimination.

According to the Federal Office of Civil Aviation in Switzerland, the guarantee of optimal connections to all important cities in Europe and around the globe is the primary goals of Switzerland's aviation policy. However, the Swiss government also strives for sustainable development and the highest safety standards in aviation. From a national perspective, the main area of conflict is the development of an attractive place for business and of an export-oriented country while needing to minimise the environmental impacts (Federal Office of Civil Aviation FOCA, 2018, pp. 8–9).

3.1 International and Supranational Air Transport Organisations

The International Civil Aviation Organization (ICAO) on a global level and the European Aviation Safety Agency (EASA) on a European level typically shape the regulations for the aviation industry outside of Switzerland. The International Air Transport Association (IATA), as a lobbying body in favour of airlines, is another air transport organisation with great influence. A short overview of those organisations can be found in Table 1 below.

ICAO aims for an efficient, safe, secure, economically sustainable and environmentally responsible civil aviation sector. Together with industry groups and the member states, ICAO agrees on policies, international civil aviation standards and recommended practices (ICAO, n.d.-a). To limit the impact of aviation emissions on local air quality and the global climate, ICAO has enacted environmental certification standards for aircraft and engine designs, which are regularly updated and tightened. In 2010, ICAO agreed on two global goals to minimise the environmental impact of international aviation: As of 2020, the industry should grow carbon

Table 1 Overview of air transport organisations ICAO, EASA, and IATA

	Role	Formation	Members
ICAO	A specialised agency of the United Nations holds legal capacity when it comes to public international law	Founded based on the Chicago Convention in 1944	193 member states
EASA	Agency of the EU, legal personality of its own	Established in 2002	31 member states
IATA	Trade association for airlines all over the world	Founded in 1945 to enable standardisation, especially concerning flight paths and pricing	290 member airlines

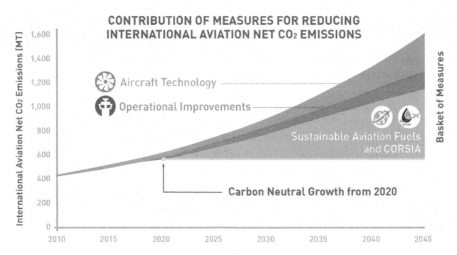

Fig. 1 Contribution of measures for reducing international aviation Net CO_2 emissions. Source: ICAO (n.d.-b)

neutral. Also, until 2050, the fuel efficiency improvement should be 2% per annum. Those targets should be achieved by sustainable aviation fuels, operational and aircraft technology improvements as well as CORSIA, the Carbon Offsetting and Reduction Scheme for International Aviation, which will be discussed later. This strategy is also known as ICAO's "basket of measure", as shown in the picture below (Fig. 1).

EASA drafts implementing rules concerning safety and environmental protection in Europe. This means that the certification and approval of products like aircraft or engines and organisations is part of their responsibility. EASA also issues Airworthiness Directives containing requirements that need to be fulfilled to operate an aircraft. The environmental standards should ensure that advanced technologies to reduce noise and emissions are part of aircraft and engine design. Equivalent to EASA but within the United States is the Federal Aviation Administration (FAA).

Even if they only oversee one nation, they may have an equally considerable influence on a global scale.

IATA has the mission "to represent, lead and serve the airline industry" (IATA, 2020). Similar to ICAO, IATA agreed on three goals to tackle climate change by reducing CO_2 emissions in aviation:

- "An average improvement in fuel efficiency of 1.5% per year from 2009 to 2020,
- a cap on net aviation CO_2 emissions from 2020 (carbon-neutral growth), and
- a reduction in net aviation CO_2 emissions of 50% by 2050, relative to 2005 levels" (IATA, n.d.).

Those targets should be reached by improving technology and infrastructure, more efficiency in aircraft operations and a global market-based measure for the aviation industry. Chapter "Introducing Sustainable Aviation Strategies" offers more insights into IATA's targets and measures.

3.2 The Chicago Convention and Air Service Agreements

To get a better understanding of aviation policy, specific agreements need to be examined as well. Concerning the operation of international scheduled flights, the Chicago Convention and air service agreements play an important role. The *Chicago Convention* was first signed in 1944, but it still contains the main principles and regulations for international civil aviation today. The Chicago Convention's goal was to create a central agreement that makes international aviation development possible in a secure and structured way and offers equal opportunities for all states. In 1971, Annex 16, "Aircraft Noise" was added as the first environmental standard. It was later renamed "Environmental Protection" since engine emissions were included (Pelsser, 2020). For *air service agreements*, negotiations typically take place on a bilateral basis between countries. Bilateral air transport agreements regulate the allocation of air traffic rights between the affected countries and their territories. Switzerland alone is part of more than 140 bilateral agreements to give Swiss carriers the chance to operate internationally. The air transport agreements, for example, regulate capacities or destinations that can be served. One of the most critical agreements was entered with the European Union (EU) in 2002 to grant Swiss airlines access to the liberalised market in Europe.

3.3 The Single European Sky Reform

When talking about the airspace system in Europe, a closer look needs to be taken at the *Single European Sky (SES)* reform. The aim of the SES reform, enacted by the EU in 2002, is to restructure the air traffic management/the airspace system in Europe to solve congestion problems and traffic inefficiencies. Furthermore, SES enables better use of the airspace, as the airspace should no longer be divided by

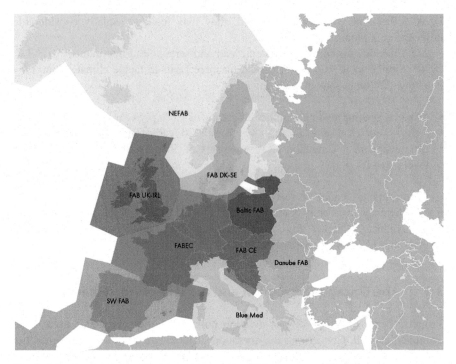

Fig. 2 Functional Airspace Blocks according to the SES programme. Own illustration

national boundaries. Currently, the fragmentation of the European sky and different Air Traffic Control technologies lead to miscoordination and mismanagement. This consumes more fuel and increases delays due to aircraft meandering between different blocks of airspace. The European airspace should be reshaped into functional airspace blocks to make integrated management of the airspace possible (Fig. 2).

The Functional Airspace Block Europe Central (FABEC) is a cornerstone of the Single European Sky. Since 2013, it covers the airspace of Belgium, France, Germany, Luxembourg, the Netherlands, and Switzerland. "The objective of the FABEC is to achieve optimal performance in the areas relating to safety, environmental sustainability, capacity, cost-efficiency, flight efficiency and military mission effectiveness, by the design of airspace and the organization of air traffic management in the airspace concerned regardless of existing boundaries" (FAB Europe Central, 2021). The Single European Sky should lead to a more effective, safer, and cost-efficient airspace and cause positive environmental effects due to shorter flight times, which lead to fewer emissions.

A reduction in flight times will be possible due to fewer delays and shorter paths. It is assumed that a 10% reduction of the environmental impact, based on 2012 levels, is possible. The first success is reducing 2.5 million tonnes of CO_2 per year since 2014 owing to a new flight procedure. However, in 2019, delays alone led to 11.6 million tonnes of excess CO_2. The detours that occur when flying through congested airspace and/or avoid charging zones with higher rates cause unnecessary CO_2 emissions. On average, each flight in European airspace takes a 49 km longer route than the direct route. As part of the SES reform, the development and implementation of new procedures have a time horizon until 2035. An upgrade of the SES regulatory framework is, for example, being discussed now, taking the European Green Deal into account. What impedes progress is that national airspace is considered sovereign territory, over which states want to retain control in the interest of national security. Furthermore, due to the current charging scheme, effective economic incentives lack the desired positive environmental impact.

4 Some General Challenges Concerning Policies for Sustainable Aviation

As the SES initiative shows, reforms do not happen quickly in the aviation industry. The design and implementation of policies for sustainable aviation face different challenges:

- *Aviation is an international industry*, different countries with different capacities and interests are included. This variety of opinions makes it hard to enforce regulatory approaches on an international level. The most influential countries can slow down or even stop the implementation of measures. Furthermore, fair access to the international market must be granted for all nations. This requires much coordination.
- Different nations and many *different stakeholders with varying opinions participate in the debate*. Part of this is a strong aviation lobby. For example, IATA's airline community may take up a different position than some member states. All stakeholders need to agree on a shared understanding of sustainability, making it hard to decide on targets. It may be possible that decisions are influenced by those wanting less strict goals.
- *Different policies exist at different levels*. New policy measures need to consider all the existing policies at a national and international level, for example, the Kyoto Protocol with common but differing responsibility among nations.
- It is easier to implement national or regional policies than international ones, but *national or regional actions could lead to unintended consequences*. For example, if national or regional policies make flying more expensive in some countries, air passengers may choose alternative routes via other countries and thus increase emissions in total. Or as democratic decision making happens either on national or even regional level, other factors such as noise are of higher priority than emissions.
- *Allocation of CO_2 emissions is tricky*. Due to airlines code-sharing or the dependence of CO_2 emissions on many different factors, it is not easy to allocate emissions from international flights to specific states.

- *There will always be resistance.* Due to government intervention failures and parties seeking private interests instead of the ones best for society, imposing prices to change behaviour and implementing subsidies for new technologies, which are more environmental-friendly, may not work. Political and public resistance needs to be expected for every measure, especially when paying for something that was free before.
- *No policy measure is perfect.* Often it is unsure what unintended consequences national or regional policies may cause. By reducing one impact, others can be increased, or new ones can appear. For example, it may be possible that the stabilisation or reduction of CO_2 emissions leads to an increase in NO_x emissions.
- *Too many measures may be implemented.* For example, if an emissions trading scheme with a fixed limit of possible emission is in place, a tax will not lead to further emission reductions. Too many measures would lead to unnecessary costs and ineffectiveness, which could end with the aviation industry's chronic loss. However, trade-offs need to be accepted to achieve the critical objectives. Therefore, it is crucial to monitor policy measures by using, for example, appropriate reporting methods.

While talking about specific policy measures targeted at making aviation more sustainable, those challenges need to be kept in mind.

5 A Policy Framework for Sustainable Aviation

The following paragraphs focus on short-term policy measures which could be implemented to reduce greenhouse gas emissions from aviation. Since the goal is to offer a holistic picture in addition to specific measures, a policy framework for sustainable aviation, which can be found on the next page, will be applied as a basis for discussion. This model was built using a grounded theory approach, based on a literature review, online (newspaper) articles, and semi-structured interviews with experts.

The policy framework differentiates the two core aspects, "demand must decline" and "remaining flights must reduce their impact on climate", which helps achieve an overall GHG reduction in aviation. To reduce the demand, four aspects were seen as particularly important: strengthen alternatives to flying, environmental education, declaration/communication of emissions, and an elaborated ticket tax. To diminish the GHG emissions from the remaining flights, a blending quota for synthetic fuels, restructuring of air space, CORSIA as a global starting point and a refined EU ETS, which will be discussed later in the chapter, were elaborated as the most critical measures that can be realistically implemented. It is essential to see that a focus needs to be put on non-CO_2-emissions like NO_x since current versions of CORSIA and EU ETS only deal with CO_2 emissions. Furthermore, the policy framework incorporates considerations for elaborating measures and considerations for politics' actions concerning sustainable aviation in general. Triggers for change, i.e., reasons why aviation must reduce GHG emissions, can also be found in the model. To better understand the framework, we will explain each part of the framework in more detail (Fig. 3).

The Role of Public Policy

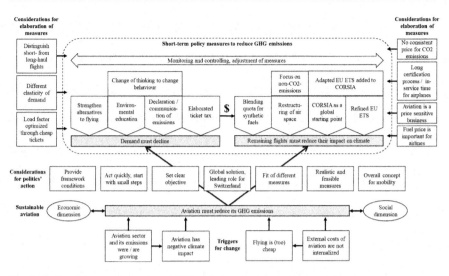

Fig. 3 Policy framework to reduce GHG emissions from aviation (own figure)

5.1 Triggers for Change and Sustainable Aviation

When looking at the framework from the bottom-up perspective, potential triggers that lead to the statement that "aviation must reduce its GHG emissions" can be found first. The aviation sector and its emissions are growing, which harms the climate. These triggers for change need to be addressed so the external costs of aviation are internalized (Fig. 4).

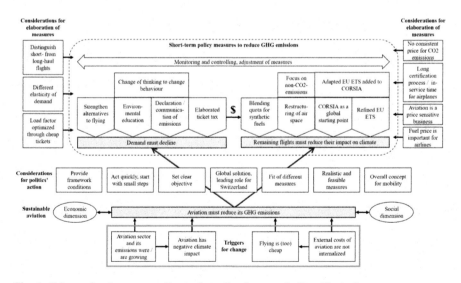

Fig. 4 Triggers for change according to the policy framework. Own illustration

As mentioned in previous chapters, when talking about sustainable aviation, the economic and social dimensions must be considered. We will not go into detail on this, but some thoughts should be presented. In general, a severe downsizing of the aviation industry is deemed unacceptable. Since Switzerland is a small, open economy and has many international headquarters, good accessibility is crucial. From an economic perspective, leisure travel is less important than business travel, and long-haul flights are more important than short-haul connections. However, the importance of incoming tourists should also not be overlooked. Concerning the social dimension, the argument of injustice caused by discriminating measures is prevalent. One can, for example, argue that access to air services will be limited for many people if flying gets much more expensive. However, it needs to be kept in mind that also, at the moment, only the world's wealthiest people are flying. When talking about global measures, it has to be kept in mind that emerging countries emit much less CO2 but will be affected to a greater extent by climate warming than countries like Switzerland (Fig. 5).

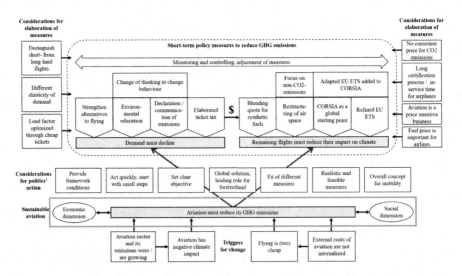

Fig. 5 Sustainable aviation according to the policy framework. Own illustration

5.2 Considerations for Political Action

If aviation intends to reduce its GHG emissions, a clear objective is needed before politics should take action. If the goal is to reach net-zero CO_2 emissions until 2050, it is questionable if CORSIA is the appropriate measure because it will not reduce emissions, but rather keep them at the same level. To reach net-zero emissions, climate-neutral flights will be needed. It is essential to have a political debate. As can be seen in the framework, another consideration for political action is the idea that countries like Switzerland must play a leading role, even if a global solution is needed.

Aviation is a global business and for this reason an ideal solution is a global one, or at least one that is harmonised at a European level. On the one hand, national measures may not yield good enough results. On the other hand, individual countries can be role models and inspire international actions. If looking at the carbon footprint per person, Switzerland has one of the highest footprints. Since not enough progress is made at a global level and there is not much time left to change the situation, the national level needs to be considered as the starting point. Countries like Switzerland should also use their power to stimulate progress at ICAO. However, measures need to be politically realistic and feasible. Each effort that aims to change the status quo will have a hard time. For example, it is not realistic that a considerable increase in ticket prices will be pushed through. Many people do not want to see flying become more expensive, which is one of the major problems. There is an attitude behaviour gap as society wants to reduce negative impacts but at the same time does not want to significantly reduce their own travel activity. Furthermore, society demands a solution where everybody is forced to reduce their travel emissions and not solely people on low incomes, who would be most affected by higher fees such as emission charges.

Furthermore, a solution must also be realistic from a technological point of view. For example, as seen in Chap. "Technology Assessment for Sustainable Aviation", it is unrealistic to implement electric flying on a large scale in the short- or medium-term. We should also aim for a coherent overall concept for mobility. The question of accessibility does not just affect one mode of transport. The climate impact of mobility as a whole needs to be solved. Different measures need to fit each other. Symbolic politics can do more harm than good to the environment, mainly because too many different measures could lead to confusion and very high administrative costs. It is crucial to carefully evaluate the measures before implementing them, as to avoid implementing measures just for the sake of taking action (Fig. 6).

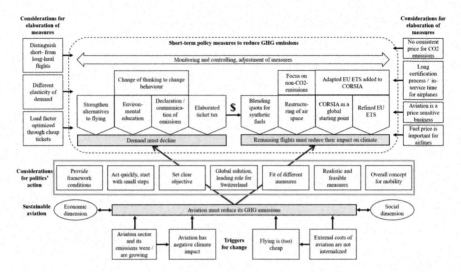

Fig. 6 Considerations for politics' action according to the policy framework. Own illustration

5.3 Short-Term Policy Measures to Reduce GHG Emissions

As explained before, the framework differentiates between measures to reduce demand and measures to reduce the climate impact of the remaining flights. It would theoretically be possible to ban air travel, but this will not happen. Especially for time-sensitive and long-distance travels, airplanes will still be used in future. Apart from a crisis like the COVID-19 pandemic, people will not fly less as long as no strict measures exist. If prices increase, people will start to change their behaviour accordingly. But this does not mean that they stop flying altogether. For example, it is possible that only long-haul flights will be avoided or that people choose to drive to another airport to start their journey from there. However, a reduction of demand does not imply a cessation of aviation emissions and thus cannot be the sole goal. Furthermore, a great part of the society may just pay the extra cost of emissions charges if they can afford it. This means that especially low-income individuals or families will fly less, while wealthy leisure travellers or business travellers will barely adjust their behaviour. Especially frequent flyers contribute most to the negative effects because frequent flyers fly often for business and are less price sensitive when they travel. The fact that they are often high earners and have the possibility to use their frequent flyer points to reduce travel costs contribute to the lower price sensitivity.

Nonetheless, demand reduction would be significant during the transition phase, where a technological solution would not exist yet. It must be taken into account that for some airlines large parts of their revenue are generated from premium classes. Business travellers are less likely to change their behaviour based on a price increase.

It is not easy to change behaviour on a voluntary basis and it is not clear what the best way to change people's behaviour is. People will need to change their way of thinking. Most important, everyone should always ask oneself if it is indispensable to travel by air. It can also be said that our understanding of mobility as a whole needs to be questioned, mainly because not everyone across the globe can travel as much as the average person in Switzerland does. The information concerning aviation and its climate impact needs to be communicated better to raise awareness. Currently, individual travel decisions depend mainly on the individual purchasing power. Even though behavioural changes take place as the phenomenon of flight shame shows, many people think that air traffic is not too bad, which indicates a lack of environmental education. This may be connected to the fact that a declaration of emissions is missing on tickets or advertisements. Therefore, transparency, which means providing information in a way that people can understand, is lacking. However, greater transparency is needed so that air passengers can make a better-informed decision. Public information campaigns will not be sufficient to reach the intended change independently, but they can help enhance other measures' legitimacy. Also, environmental education beginning at an early stage in education is necessary.

To support people's change in their travel behaviour, it is important to have alternatives to flying that can provide the connectivity needed. Rail services are an alternative which can be improved, mainly to substitute short-haul flights (e.g. night train offers, better high speed train connectivity, better punctuality on a European scale, better timing and scheduling to reach places in the early morning and get back the same day in the evening, etc.). A good example is the TGV Lyria connection from Switzerland to Paris, which makes flying to Paris less attractive. Even if short-haul flights' substitution does not significantly impact the substitution of long-haul flights, it needs to be kept in mind that around 3/4 of local passengers departing from Switzerland have their final destination in Europe. However, there is not enough capacity to completely transfer the air traffic onto the railways. Chapter "Introducing Sustainable Aviation Strategies" offers more insights into the current European train network. Besides, it needs to be kept in mind that building rail infrastructure also leads to environmental consequences. Road transport as an alternative to flying, in general, should not be fostered as it may generate even higher GHG. However, buses for short-haul routes may be a valid alternative.

One possibility to make flying more expensive is a ticket tax/air passenger tax. Different countries have already imposed departure taxes. Some taxes vary depending on the distance that is flown, others were set at a fixed amount. Most of the time, the money earned through those taxes goes directly to the government budget. Some countries had similar taxes some time ago but lifted them in the meantime. One example is Denmark which mentioned the loss of customers to Swedish airports as the reason to abandon the tax in the year 2006. As discussed in Switzerland, the ticket tax is an incentive fee that should lead to less flying. It is not the goal to increase the country's tax revenue. It is questionable if such a tax will change people's behaviour. The proposed ticket tax in Switzerland would range between CHF 30 and 120 and would be mandatory for all passengers departing from a Swiss airport. The quantification of a future decline in demand due to this tax is contentious.

An economic analysis by the EPFL and University of Lausanne shows that, depending on the final levy, a CO_2 reduction of up to 20% (compared to 2018) may be possible. However, if CHF 30 needs to be paid per economy and CHF 60 per business class ticket for short-haul flights and CHF 90 in economy and CHF 120 in business for long-haul flights, a reduction of 7–11% is seen as likely (Fumagalli, 2020). Airlines operate in oligopolistic but dynamic markets with fixed frequencies based on a limited number of airport slots which are a scarce resource. For this reason, airlines tend to look at the total cost of emissions per flight and not at a passenger level. They can cover these emissions costs either by reducing other costs or increasing certain price components for some passengers. The latter would imply that last-minute customers, who are less price sensitive, may pay more for emissions than early bookers, who may not pay at all. It would also be possible that emissions charges are distributed as small mark-ups on different ancillaries such as baggage fees, seat reservation, credit card fees, food sales in planes, etc. It is also a fact that airlines entice passengers with low anchor prices in their marketing and by doing so initiate a thinking process with potential travellers. By the time the traveller books, prices have already increased, but as an implicit travel decision has already been made subconsciously, the willingness to pay increases and extra charges such as emissions charges have less of an impact.

Therefore, it is argued that there will be no steering effect with a ticket tax that only charges CHF 30 and in the absence of an exact definition on the implementation of the tax. A report on the implementation of the German passenger charge (Infras 2011) has shown that it only resulted in a small reduction of demand growth for a short period of time, hence, aviation's growth will not be stopped entirely through such a measures. Thus, one of the most important questions is if and how airlines pass on the ticket tax to passengers. Airlines could bear the additional costs themselves in exchange for a better load factor. It is also criticised that Switzerland's neighbouring countries have lower ticket taxes than what Switzerland is currently evaluating. This could lead to passengers departing from neighbouring countries to get a cheaper connection. If much more is charged for long-haul than short-haul flights, people may choose to take a short-haul flight to another country and take the long-haul flight from there instead of Switzerland to save money. This would have a larger negative impact on the environment than if they would have flown directly from Switzerland to their destination. There will not be a consistent European ticket tax in the near future as not all countries would agree. One advantage of a national ticket tax is the possibility to combine it with other European or international measures. A point for discussion is the use of the money gained collected from taxation. The money should be invested in climate protection and new technologies and fuels, instead of being redistributed among the population, to provide future proof and long-lasting solutions. Money could be spent on new technologies and synthetic fuels for aviation and other climate protection projects like overnight trains. Investing in projects that have a positive climate impact seems to be essential.

As seen in Chap. "Technology Assessment for Sustainable Aviation", synthetic fuel could be a solution, at least at an interim stage. Especially for long-haul flights that are hard to substitute, it looks like synthetic fuels would be the best option. Solar fuels may be preferable since they have a higher degree of efficiency than power-to-liquid fuels. However, non-CO_2-effects will not disappear entirely as long as common jet-engines are used. An essential advantage of synthetic fuel compared to biofuel is that no agricultural land must be used since solar panels could also be built in the desert. However, it needs to be kept in mind that significant infrastructure would be needed, which takes years to be built. Even if only 1% of what is fuelled in Switzerland would be synthetic fuel, already a considerable size of infrastructure is needed. It is estimated that it would take approximately 40,000 km^2 of solar panels in sunny countries to produce enough solar fuel to meet aviation's global demand. It could take at least 20 years to build such infrastructure. However, a first plant (approximately 10 km^2) should cover 2% of the fuel used for aviation in Switzerland. With the money earned through the ticket tax, it would be possible to fund such a plant. Airlines cannot bear the additional costs of synthetic fuels since they are much more expensive than fossil fuels. Production costs will only decrease if larger amounts are produced. For renewables energies, costs usually decrease by approximately 15–20% when the infrastructure is doubled. For synthetic fuels to be competitive, they need to become cheaper. Different possible solutions to fund synthetic fuels exist. We should learn from examples like photovoltaics development where a feed-in remuneration at cost was very successful. At the moment, the security of investment for synthetic fuel is missing. For the first quantities, the certainty of demand is needed. It would also be possible to help production facilities with subsidies or direct investments. To get the industry and research onboard, public–private partnerships are needed. Kerosene is hard to replace due to its characteristics for flight safety and its high energy content per kilogram.

Since sustainable aviation fuel faces some challenges, policy support is needed to push it. One possibility is to enact a quota ensuring that a certain percentage of aviation fuel used is not kerosene. Starting in 2020, a biofuel quota exists in Norway. At least 0.5% of the aviation fuel used by airlines that operate in Norway needs to be advanced biofuel. Therefore, airlines must make use of this fuel even if it is more costly. By 2030, 30% of the jet fuel should come from second-generation biofuels. This means that only products based on waste and leftovers are allowed, so the use of palm oil is prohibited. A mandatory blending quota for synthetic fuel would also be a possible accelerator for synthetic fuel since it would ensure demand. If starting in 2025, 1% of what is fuelled in Switzerland must be synthetic fuel, it would be possible to raise this quota to 100% by 2050. It would be possible to agree on a partial coverage of the extra costs associated with the use of synthetic fuel using the revenue from a ticket tax, for as long as synthetic fuel is considerably more expensive than fossil fuels. The goal should be to use synthetic fuels without creating a substantial competitive disadvantage for airlines using it. Tankering could be one disadvantage, if a substantial price difference exists compared to other airports, due to Switzerland having a synthetic fuel quota whereas other European countries would not. Therefore, a European solution would be desirable. It also has to be kept in mind that other sectors like the automotive industry would like to benefit from synthetic fuels.

As stated before, synthetic fuels could help to reduce non-CO_2-effects of aviation. Those need to be targeted as well when talking about policy measures. Mostly because the current measures CORSIA and EU ETS (the European Union Emission Trading Scheme) for aviation are only dealing with CO_2 emissions, no incentive to reduce non-CO_2-emissions exists. This means that the further development of engines targets a reduction in CO_2 emissions and not in other emissions. If demand for flying decreases, non-CO_2-emissions decrease too. However, to put non-CO_2-emissions more into focus, it would be possible to restructure the air space and optimise the routes. One possibility is tactical routing based on the weather situation to get a minimal overall climate effect, for example, with restricted climate airspaces. The idea is to have weather-optimised flight trajectories that lead to a minimisation of emissions in climate-sensitive regions. Since they have a short lifespan, the climate impact of non-CO_2-emissions depends on the time and location of the emission. In other words, they are primarily dependent on meteorological and chemical background conditions. Having restricted climate airspaces means that regions that are highly climate-sensitive at a certain point of time are closed for several hours up to several weeks, and flights are re-routed in order not to cross them. To decide which airspace is restricted and cleared, policymakers need to set a threshold value for a flight's climate change contribution. The implementation of climate-restricted airspaces can be done by air traffic control. To prevent a capacity crunch, only the most ecologically harmful trajectories should be closed. Instead of closing climate-sensitive regions, it would also be possible to have different charges for flying through them (Niklass, et al., 2019, pp. 102–103, 108–109). However, when looking at the lengthy and slow progress of the Single European Sky reform, it is clear that it is hard to achieve a restructuring of the airspace.

When coming back to measures that deal with the CO_2 emissions in aviation, CORSIA seems to be a controversial one. CORSIA is a global market-based offsetting scheme that enables the compensation of CO_2 emissions to reach neutral growth in international aviation's CO_2 emissions from 2020 onwards. Every year starting from 2019, airplane operators from ICAO's member states that conduct international flights have to monitor, report, and verify their CO_2 emissions. This applies to all international flights, even if a state is an ICAO member but not part of CORSIA. The pilot phase of CORSIA is set for the years 2021–2023.

In the first phase (years 2024–2026), states' participation is voluntary. During the second phase (years 2027–2035), participation is mandatory for all states that accounted for an individual share of more than 0.5% of total international aviation activity (in RTKs) in 2018 and for all states whose cumulative share reaches 90% of total RTKs. Not included are the least developed countries, small island developing states, and landlocked developing countries. CORSIA is a route-based approach. This means that an international flight is covered by CORSIA as soon as the origin and the destination of it are in a country participating in CORSIA, indifferent of the aircraft operator. Equivalent to the total of the final offsetting requirements, which is calculated every year, the airplane operators need to purchase and compensate eligible emissions units from the global carbon market. The offsetting requirement can be reduced by using sustainable aviation fuels.

CORSIA is essential because it is a global measure with the same rules for everyone, but it is not a perfect solution. What is criticised the most is that all emissions up to the year 2020 levels are free of charge and that participation is voluntary until 2027. Therefore, it is difficult to say what will be achieved by implementing CORSIA. The large decrease in flight traffic due to COVID-19 may lead to the CORSIA objective being met without doing anything. However, aiming for carbon-neutral growth will not be enough if net-zero wants to be reached till 2050. Some open questions concerning the offsetting process remain unanswered as well. It is, for example, criticised that already existing emission allowances can be used for offsetting too. In general, improvements could be made when it comes to the credibility of emission allowances. Buying offsets is way cheaper than, for example, investing in synthetic fuels. As stated before, CORSIA does not cover aviation's non-CO_2-emissions and therefore ignores at least half of aviation's climate impact. However, one advantage of CORSIA is that CO_2 emissions from aviation are measured. This could lead to having CO_2 emissions as a new key performance indicator. Another opinion exists that CORSIA is a measure that should be pursued because it is the market-based measure ICAO decided upon, and therefore the only global one that is possible at the moment. Since all critical players are participating, airlines' costs for CORSIA will be reflected in ticket prices. However, CORSIA should not be seen as an excuse to not implement other measures.

It has not been decided yet if aviation will still form a part of the EU ETS when CORSIA has been fully implemented. It will probably only be possible to have both measures if the EU ETS is adapted. It would, for example, be possible that CORSIA deals with the emissions above the 2020 level while the EU ETS looks at the remaining emissions. Thereby, more than "just" carbon-neutral growth could be achieved. Another possibility would be to have the EU ETS in place for domestic flights not included in CORSIA.

The EU ETS is seen as the critical tool of the EU's policy to reduce GHG emissions cost-effectively among different sectors. Since 2012, CO_2 emissions, but not all GHG emissions, from commercial aviation are part of the EU ETS. A refined EU ETS could also include non-CO_2-effects. The idea of an additional submission of allowances for NO_x emissions or H_2O is something that needs to be kept in mind when refining the EU ETS. Due to resistance from countries outside the EU, the geographic scope is limited to flights within the European Economic Area, consisting of the EU member states plus Iceland, Liechtenstein, and Norway, until the end of 2023. The airlines conducting such flights (some exemptions apply, for example, for flights performed under public service obligations) are required to monitor, report, and verify their emissions. After each year, they must surrender enough allowances to cover their emissions. The aircraft operators receive free emission allowances, which can also be traded on the market, covering a certain level of emissions from their flights, based on a benchmark established in 2011. For all emissions above this level, the operators need to buy allowances on the market. It is possible to buy allowances from other industries that are part of the EU ETS, which means allowances do not need to be generated in the aviation sector. The price for those allowances is not set but determined through trading (demand and supply). The revenue from auctioning goes to the member states. At least 50% of this money gained should then be spent on reducing climate change, developing renewable energy, or investing in R&D. The EU ETS works as a "cap and trade" system. Thereby, the cap (total amount of GHGs that can be emitted) is reduced over time. For the year 2012, the cap equalled to 97% of historical emissions (the mean of the years 2004–2006), for the years 2013–2020, the cap is set to 95%.

Switzerland and the EU's agreement to link their emission trading schemes went into force on the 1st of January 2020. Most aircraft operators offering domestic flights or flights from Switzerland to the EEA are now required to participate in the Swiss ETS. Free of charge emission allowances for aircraft operators are calculated based on their transport performance in 2018 and a benchmark. A certain amount of emission allowances will be placed on auction. For flights from the EEA to Switzerland, the EU ETS applies.

The EU ETS was often criticised because its effect was not as considerable as expected during the last years. Too many emission allowances were given away for free and prices were lower than initially calculated. It has to be taken into account that the EU ETS includes the aviation sector and other ones like electricity production. This means that an emission reduction does not have to occur within the aviation sector but that emissions are reduced where it is the cheapest to reduce them. With a price for an emission allowance of around 20 to 30 Euros, airlines have no incentive to invest in synthetic fuels. With such prices, kerosene's price fluctuation makes a more significant difference than what needs to be paid due to the EU ETS.

Furthermore, the limitation to flights within Europe is also seen as unfavourable. However, for the EU, the EU ETS is expedient instead of individual measures. It is seen as positive that it is a European measure affecting more than just one country. It is expected that a larger part of the emission allowances needs to be bought in the future.

When looking at the different policy measures to reduce GHG emissions from aviation, it is crucial to do proper monitoring and controlling to adjust and reach the defined objectives. For example, it would be possible to increase the ticket tax. This is especially important since it is impossible to know all consequences that a particular decision will have in a business that is global and very complex like aviation (Fig. 7).

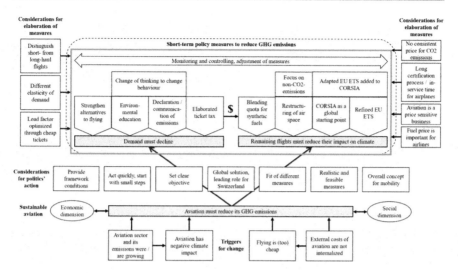

Fig. 7 Short-term policy measures to reduce GHG emissions according to the policy framework. Own illustration

5.4 Considerations for Elaboration of Measures

When elaborating the measures, some considerations need to be taken into account. The first one is the distinction of short- and long-haul flights. The largest part of aviation's emissions is generated by flights with a length of more than 1500 km. The decarbonisation of those flights requires special attention if emission reduction should take place on a large scale. However, those flights are also the ones that cannot be substituted easily, for example, by the usage of other means of transport such as trains. Therefore, it can be concluded that long-distance flights will rely on synthetic fuels to reduce their climate impact in particular because new technologies like electric flying are more realistic for short-haul flights. As stated before, policy measures, depending on how they are implemented, can have different effects on short- and long-haul flights. It is, for example, expected that a ticket tax would have a more significant effect on short- than long-haul travels.

Those different effects can be explained when looking at the elasticity of demand. If prices change, demand will change accordingly. However, it is difficult to forecast specific changes in demand due to behavioural and dynamic pricing issues. As said before, despite differentiating between short- and long-haul flights, other aspects such as the purpose of travel (leisure or business) and the class (economy, business, first) need to be looked at too.

When implementing measures that should make flying more expensive, it must also be taken into account that airlines will always try to optimise their load factor. An airplane has a large portion of fixed costs. Those fixed costs per seat decrease when the load factor increases. Therefore, cheap tickets are offered to get a better load factor.

Aviation is a price-sensitive business. Airlines optimise their costs, not their CO_2 emissions. If profits are not high enough, an airline will disappear from the market. Airplanes may, for example, detour when air traffic control charges are, as an example, higher in Italy than in Croatia. Airlines also calculate exactly where to refuel how much, always taking the weight of the plane into account, as a significant fuel burn is used for carrying the weight of the fuel itself.

The amount paid for fuel is essential for airlines since fuel is responsible for a great part of operational fixed costs. However, this can positively affect aviation's climate impact since possibilities to increase efficiency are implemented. New airplanes have considerable advantages when it comes to fuel consumption. This can influence the procurement of new airplanes.

However, experts often mention that airplanes have a long in-service time which means that airplanes may be in use for decades. It has shown that high-quality Network airlines often have an average fleet age of about 10 years. Some airlines keep the average age even below 10 years. But once an airline does not use planes anymore, it sells them to other carriers or reconfigures them as cargo aircraft where they stay airworthy. Furthermore, it takes up to 20 years until airplane producers such as Airbus and Boeing reach a break-even for their investments into production lines as well as airplane innovation and engineering. This makes it especially hard to implement completely new technologies like hydrogen-powered aircraft. Another point to keep in mind is that it may take many years until a new airplane is certified and can enter the operational market.

The final consideration for the elaboration of measures presented in the framework is the fact that no consistent price for CO_2 emissions currently exists. Depending on the study, the costs per tonne of CO_2 vary greatly. An estimation of costs is challenging because the effect of CO_2 emissions can last up to approximately 100 years at high altitudes, whereas it can disappear quicker at ground level (Bundesamt für Zivilluftfahrt BAZL, 2020). A global price signal for CO_2 would be needed to create demand for new technologies. Only then it would be possible to calculate how much can be saved by using synthetic fuel instead of fossil fuel. Currently, the external costs of aviation are unknown. However, it is clear that aviation's external costs need to be internalised and that this can be achieved using different means (Fig. 8).

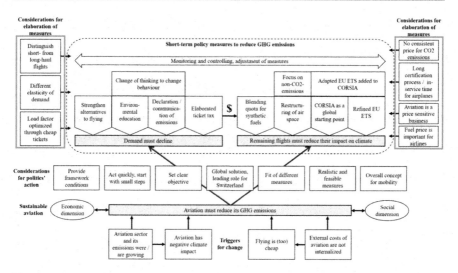

Fig. 8 Considerations for elaboration of measures according to the policy framework. Own illustration

5.5 Currently Unwanted or Unrealistic Policy Measures

Of course, there are more possible policy measures than the ones already discussed. However, many of them are deemed unrealistic or unwanted. For example, a sought-after but unrealistic measure is the international taxation of kerosene. Fuel taxation is one of the most discussed ideas of environmental charges, since it would enable the implementation of the "polluter pays" principle on a global level. Fossil fuel would become more expensive, which leads to alternative fuels and technologies becoming more attractive. The final impact of a tax on aviation fuel is hard to quantify since the effects of the tax is unknown to a certain degree. For example, it cannot be taken for granted that such a tax would lead to effective emission reductions if alternative fuels were not available. The impact of an aviation fuel tax also depends on how airlines handle the additional costs. If consumers are too price-sensitive, airlines will need to factor the extra costs into their cost structure.

In contrast, if the costs are passed onto consumers and ticket prices increase, reduced demand for air travel would lead to lower emissions. However, a tax on fossil fuel is currently not capable of winning a majority internationally. It would be hard to implement since many air service agreements forbid levies or taxes on fuel in transit or exempt fuel that is loaded into an aircraft in the other state from taxation. For example, the EU-US Open Skies Agreement exempts fuel for international air transportation use from taxation based on reciprocity (The Council of the European Union, 2007). Thus, a US carrier that is conducting flights between two EU countries is exempted from fuel taxation. To introduce an aviation fuel tax, such agreements would need to be re-negotiated. Since all affected countries would need to agree, the implementation would be slow.

Another problem that the implementation faces is the fact that kerosene is not used where it is fuelled. If aviation fuel is more expensive in a particular country due to tax, it may be possible that airlines start to refuel in another country. This would lead to heavier aircraft and more emissions. Taxation only at a national level also leads to a competitive disadvantage and is, therefore, less desirable.

Flying could also become more expensive if flights were subject to Value Added Tax. In Switzerland and most EU member states, domestic flights are already subject to Value Added Tax. By contrast, flights between EU member states and international flights are not subject to Value Added Tax. For example, the EU VAT Directive states that Value Added Tax cannot be charged for international flights. Since international aviation operates in a multiplicity of jurisdictions with different Value Added Tax rates, equitable treatment would be hard. If, for example, Switzerland would like to levy Value Added Tax for international flights, it would only be possible to charge those parts of a flight above Swiss territory, leading to high administrative expenses.

A minimum price per flight/a ban of low-cost flights is again a desired measure that is probably hard to implement. Austria implemented minimum prices for very short flights from Vienna, but airlines are challenging this decision in European courts. Due to the lower prices, people started to fly more often and more considerable distances. A desirable effect of a minimum price would be that journeys would see an increased price level and that the airlines' dynamic pricing would be stopped or shift to a higher level, from which it would still be possible. It can be argued that a minimum price would cause fewer journeys because frequent flying quickly became a status symbol thanks to cheap flights. Studies showed that increased ticket prices have an impact on people's behaviour. However, it would be hard to set a specific limit as a minimum price since the significance of the steering effect is unclear. It is also possible that the expensive tickets could become cheaper with a minimum price as airlines would still want to optimise their load factor. If the goal would be to have a minimum price for the whole of Europe, setting a limit would be even more difficult. Inside Europe, different countries have different levels of average earnings and assets. Furthermore, it has to be kept in mind that a minimum price does not lead to a long-term solution since no incentives for new technologies or synthetic fuels are generated.

A controversial measure that faces various difficulties is offsetting. Voluntary carbon offsetting can be conducted by airlines or customers. Since November 2019, easyJet offsets the carbon emissions caused by fuel for every flight. Swiss International Air Lines offers its customers the possibility to pay for the flight's CO_2 emissions during the booking process. As seen for CORSIA, a first disadvantage mentioned is that a lot of offsetting projects would have taken place anyway, instead of offering new offsets. A reduction of emissions may, for example, be achieved by financing an extension of forestry. It is a problem that offsetting projects are hard to control, mainly because most projects occur in developing countries. Therefore, it is hard to calculate the exact amount of emissions that can be used for offsetting.

Another problem are individuals who clear their guilty conscience by offsetting their flights, despite them not having actually reduced their emissions. Therefore, it is problematic when offsetting justifies flying. Emissions should be reduced and not just offset. However, offsetting can be a starting point in the absence of other solutions.

6 Conclusion

In the long-term, the goal must be to reach climate-neutral flying. Since this is not technologically possible within the time frame required, policy measures need to be implemented to reduce GHG emissions from aviation in the short-term. Measures to reduce demand and lower the climate impact of the remaining flights are required. Aviation is an international business and therefore it should aim to create a global solution. CORSIA is a response to the calls for a global solution. However, it is not deemed the perfect solution and should not be an excuse to abstain from implementing other measures. For example, it is crucial to put a focus on non-CO_2-emissions such as NO_x since the current version of CORSIA only deals with CO_2 emissions. Countries like Switzerland should play a leading role and implement national policies, mostly because limited time is left to prevent severe negative effects of climate warming. A change of thought will be required to reduce demand. Greater transparency concerning aviation's GHG emissions is needed to enable customers to make a well-informed decision. For example, emissions should be declared on tickets and in advertising. Another essential part is environmental education. More attractive and competitive alternatives to flying could reduce the demand for air travel. To reduce the GHG emissions of flights still operate, synthetic fuels are a promising solution. Different possibilities exist to progress with synthetic fuels. For instance, it would be possible to use the money earned from a ticket tax to invest in the development of synthetic fuels. Furthermore, a mandatory blending quota would ensure a sustained demand for synthetic fuels.

References

Bundesamt für Zivilluftfahrt BAZL. (2020). *CO_2-Emissionen des Luftverkehrs. Grundsätzliches und Zahlen*. Von Bundesamt für Zivilluftfahrt: https://www.bazl.admin.ch/bazl/de/home/politik/umwelt/luftfahrt-und-klimaerwaermung.htmlabgerufen

FAB Europe Central. (2021). *FABEC - a cornerstone of the single European sky*. Von FABEC. Retrieved from http://www.fabec.eu/about/abgerufen.

Federal Office of Civil Aviation FOCA. (2018, August). *ICAO action plan on CO_2 emission reduction of Switzerland*. Retrieved from ICAO Aktionsplan zur CO_2 Reduktion der Schweizer Luftfahrt https://www.bazl.admin.ch/bazl/de/home/fachleute/regulation-und-grundlagen/umwelt/icao-aktionsplan-zur-reduktion-von-CO_2-emissionen-der-schweizer-.html

Fumagalli, A. (2020). *So wirksam ist die geplante Lenkungsabgabe von 30 bis 120 Franken auf Flugtickets*. Von Neue Zürcher Zeitung. Retrieved from https://www.nzz.ch/schweiz/so-wirksam-ist-die-flugticketabgabe-von-30-bis-120-franken-ld.1545724?reduced=trueabgerufen

IATA. (2020). *About us*. Von IATA. Retrieved from https://www.iata.org/en/about/abgerufen

IATA. (n.d.). *Climate change*. Retrieved from IATA: https://www.iata.org/policy/environment/Pages/climate-change.aspx

ICAO. (n.d.-a). *About ICAO*. Retrieved from ICAO https://www.icao.int/about-icao/Pages/default.aspx

ICAO. (n.d.-b). *Climate change*. Von ICAO. Retrieved from https://www.icao.int/environmental-protection/pages/climate-change.aspxabgerufen

Niklass, M., Lührs, B., Grewe, V., Dahlmann, K., Luchkova, T., Linke, F., & Gollnick, V. (2019). Potential to reduce the climate impact of aviation by climate restricted airspaces. *Transport Policy, 83*, 102–110.

Pelsser, A. (2020, January 10). *The postal history of ICAO*. Von ICAO. Retrieved from https://applications.icao.int/postalhistory/annex_16_environmental_protection.htmabgerufen

Peter, M., Zandinella, R., & Maibacg, M. (2011). Volkswirtschaftliche Bedeutung der Zivilluftfahrt in der Schweiz. INFRAS 2011.

The Council of the European Union. (2007, April 25). *2007/339/EC: Decision of the Council and the Representatives of the Governments of the Member States of the European Union, meeting with the Council*. Von EUR-LEX. Retrieved from https://eur-lex.europa.eu/legal-content/EN/TXT/?uri=CELEX:32007D0339abgerufen

UNFCCC. (n.d.). *Kyoto protocol - targets for the first commitment period*. Retrieved from United Nations Climate Change https://unfccc.int/process-and-meetings/the-kyoto-protocol/what-is-the-kyoto-protocol/kyoto-protocol-targets-for-the-first-commitment-period

United Nations. (1992). *United Nations Framework Convention on Climate Change*. Retrieved from UNFCCC https://unfccc.int/files/essential_background/background_publications_htmlpdf/application/pdf/conveng.pdf

Towards Sustainable Aviation: Implications for Practice

Adrian Müller, Alexander Stauch, Judith L. Walls, and Andreas Wittmer

Abstract

- The time to act is now. Even in the recovery phase after COVID-19, the industry cannot afford to further postpone the transition to more sustainable operations. Immediate measures and long-term initiatives go hand in hand.
- The aviation industry should not see becoming more environmentally sustainable primarily as a cost factor, but as a strategic opportunity that is also financially viable in the future.
- Aviation stakeholders should be open to far-reaching change at an early stage, as this will happen sooner or later anyway. First-movers will have advantages over laggards.
- Ambitious and coordinated efforts are needed from all key players in the aviation system to drive the decarbonisation of aviation fast enough and far enough.

1 Sustainable Aviation: Key Takeaways

Aviation is an important element of our established socio-economic structures. With advantages such as a significant contribution to global economic performance and connecting different cultures, a continuation of the growth trend in aviation seems desirable at first sight. However, when considering the parallel growth in greenhouse gas emissions from burning fossil fuels and their contribution to climate change, it becomes obvious that a continuation of the status quo is not an option and has far-reaching consequences well beyond the boundaries of the industry. Aviation must mitigate its negative impact on the climate.

In this book, we have looked at and analysed the topic of sustainable aviation from different angles. In this concluding chapter, we would first like to revisit the main messages of our book and, second, offer concrete recommendations and guidance for the various stakeholders in the industry to achieve the goal of sustainable aviation.

1.1 Drive the Change: Seize Strategic Opportunities

Our first and probably most important message to aviation stakeholders is that the transition to sustainable aviation is primarily a great strategic opportunity. Climate change *will* bring about changes—both positive and negative. If we do not act, the negative consequences will be so severe that they will far outweigh the positive benefits of aviation, and even fundamentally undermine the legitimacy of the industry. For strategically far-sighted players, however, the benefits of first-mover advantages supersede the costs of investing in new technologies and processes while outweighing the cost of dealing with the consequences of climate change (see Chap. "Introducing Sustainable Aviation Strategies"). However, this only applies to those companies that, despite the difficult post-pandemic market environment, set the course now for a sustainable future in aviation and do not wait for their competitors to act or for the regulator to oblige them to act.

1.2 Act Now: Immediate Measures and Long-Term Visions

Timely action is essential not only from a strategic perspective, but especially because of the clear timeline for acting on climate change. While the industry goal set by IATA is to halve CO_2 emissions, compared to 2019 levels, by 2050, we argue that a more ambitious target is necessary and the industry should strive to become climate neutral by 2050. Considering the long innovation and approval cycles, strategic decisions and investments must be made now. Measures already available, such as fleet renewal, sustainable aviation fuels, behavioural nudges, or offsets, can be implemented in the short term. At the same time, research into new technologies (hybrid electric aircraft, hydrogen, new aerodynamic concepts), processes, and business models must be driven forward.

1.3 Coordinate the Mitigations: Systemic Interdependencies and Prisoners Dilemma

Among European airlines, in particular, there are increasing calls for a "level playing field," which is understood to mean global regulatory equality for all players when it comes to climate protection measures. The reason for these demands is the fear that unilateral local regulation—for example, at the European level—can lead to considerable price and competitive disadvantages for local airlines compared to peers from Asia and other parts of the world. Does this fear sound familiar? Yes, airlines believe they are trapped in a prisoner's dilemma.

The Prisoner's Dilemma is one of the best-known mathematical games from game theory, which is often used to describe many economic (but also non-economic) situations, in which the player who moves (first) to collaborate experiences a disadvantage, and the incentive is to opt for a lose–lose outcome in which players do not cooperate. International relations and strategic decisions of individual economic actors can also be explained by the prisoner's dilemma. In a simple view, this standard model is sometimes used to analyse why airlines are so resistant to taking climate protection measures, in comparison to other industries.

Each airline pursues its own strategy or its own interests in order to create revenue for itself and to position itself better vis-à-vis its competitors. Each airline is therefore faced with the same choice: does it cooperate with the other airlines to jointly contribute to environmental protection or does it opt to pursue its own interests? In this decision, the benefit or profit each airline receives from its driven strategy also plays an essential role.

If an airline decides not to contribute to sustainable aviation and thus not to implement the necessary measures to avert global warming, "the player" is always better off in the short term (cost savings, lower prices), regardless of what strategies the other competitors choose. In the best case, the player receives a "profit". This situation occurs precisely when the airline, as a free rider, profits from the environmental protection of the other players and makes no contribution itself.

In the case of airlines and climate change, however, the model is flawed. Although airlines may benefit from their non-cooperation in the short term, the medium to long-term consequences for them is much greater, as Chap. "Airline Perspective" has shown in detail. For this reason, all stakeholders should opt for a cooperative strategy, which means taking joint action and launching strategic initiatives together. Moreover, due to the systemic interdependencies that have also been pointed out, we call on the actors to consider their measures not in isolation but in the entire aviation system and to act together. Joint and collaborative strategies are commonly seen as the only way to tackle sustainability problems across different industries. Section 2 in this chapter shows how the key players influence each other and what that means for the key actions to be taken.

1.4 Be a Lighthouse Industry: Technological Leadership Beyond Industry Borders

A common argument is that aviation's contribution to climate change is too small to make a meaningful impact. We counter that aviation has played a special role in our technological progress since the very beginning and has contributed significantly to upstream and downstream innovation. Moreover, aviation's contribution to climate change is bigger than the contribution of the economies of Canada and Germany and yet is still growing strongly. Last but not least, the impact of aircraft's GHG emissions at high altitudes is worse than in other industries with operations closer to sea level.

More than ever, aviation can play a leading role in societal progress. Whether it is through its tradition of engineering in the development of technologies for decarbonisation, or as an industry with a long history of enabling innovation in customer operations. We believe if aviation can achieve decarbonisation, it is possible in all sectors. As a lighthouse industry, aviation can once again prove how central it is to socio-economic progress. The following figure (Fig. 1) summarises the key messages.

2 Recommendations and Guidance for Actors in the Aviation System

In this book, we take an in-depth look at technology and four actors in the aviation system: consumers, policymakers, airlines, and airports. In this chapter, we now not only want to propose very specific recommendations and advice for each of these actors, but also revisit how they interact in the context of sustainable aviation.

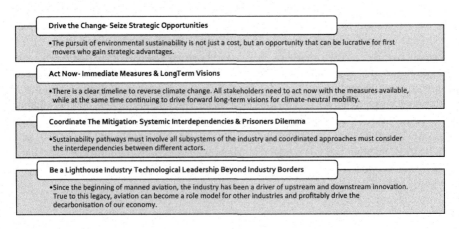

Fig. 1 Key messages

2.1 Technology

In the field of sustainable aviation, great hopes are pinned on technological progress. Our discussion of the most promising alternatives and possible development scenarios shows that innovations such as new fuels, aircraft concepts, and propulsion systems will play an important role in the short run but will not be sufficient in isolation to decarbonise aviation. However, technology is the foundation for many of the measures and strategies we address in the other subsystems.

More specifically, in the short term, synthetic fuels, lightweight technologies, and more efficient engines are important to move closer to the goal of carbon neutrality. In the long run, battery or hydrogen–electric concepts are a possible way forward to electrify short-haul (battery/hybrid) and long-haul (hydrogen) flights. Because of the inherent uncertainty in innovation, it is not clear whether and how the more disruptive concepts can be implemented. The availability of low-carbon innovations remains highly dependent on the investments of large industry incumbents (Airbus, Boeing), new start-ups but also on the development of other industries (regarding hydrogen and battery).

Because the time frame to achieve climate neutrality in the industry is short, it is crucial to understand that waiting for radical technological change to solve the problem is not viable. With current and foreseeable technologies, we can improve the impact on climate change, but not solve it to the extent that it needs to be. In addition, technology is not a silver bullet that will make aviation carbon-neutral quickly. This must be taken into account when we discuss developments in the remaining subsystems, otherwise, there is a danger that the "technology myth" as a solution will obstruct the necessary change in other areas.

2.2 Consumers

From our perspective, the customer is critical in the journey to sustainable aviation for several reasons. Most importantly, the customer ultimately determines the type and quantity of demand. A fundamental shift in mobility is only conceivable if something changes in people's travel behaviour. The first focus is on the quantity and distance of travel. Less frequent trips automatically reduce per capita emissions, shorter distances enable a modal switch, for example to rail, which may further reduce the CO_2 emitted per kilometre. Furthermore, it is in the power of the customers to dictate what kind of offers they prefer or does not want. The willingness to pay for the scope of service, for sustainable fuels or offsets also has an influence on the decarbonisation of aviation. As a result, they have a significant influence on the airlines.

However, as demonstrated by the example of climate strikes, consumers also have an influence on society as a whole. Activism such as Fridays4Future can put pressure on legislators and result in additional regulations such as airline ticket levies or bans. All consumers are influenced or controlled in their behaviour by-laws either directly or indirectly via price measures. On the road to sustainable aviation, political discourse is key and consumer activism can have far-reaching positive or negative effects on efforts to decarbonise the sector.

For us, it is therefore clear that climate-conscious consumers can neither pass the buck to the airlines nor look for excuses in technological progress. Changes in consumer behaviour can have a meaningful impact on reducing emissions. Consumers should be aware that their choices regarding travel have a large impact on their environmental footprint. Policymakers, but also airlines, can increase this awareness through information campaigns or policy interventions and steer consumer behaviour by offering alternatives to flying (e.g. high-speed rail connections). But finally, it should be pointed out that, at least in democratic systems, it is up to the population to demand political change and we must understand that consumer choices are one of the most influential pieces of the puzzle in decarbonising the transport sector.

2.3 Policymakers

As we have just seen, there is a close connection between consumers and policymakers. By its very nature, state sovereignty makes the legislator the most powerful and therefore central actor. Aviation-specific regulations on national or supranational (ICAO, EASA etc.) level have a great influence on airlines and airports, examples being blending quotas, curfews, and other operational requirements, while general laws and taxes can steer demand and behaviour of consumers, if implemented correctly.

With regards to environmental externalities, aviation is still little regulated compared to other industries. The reason is the internationality of aviation. As mentioned earlier, playing fields may not be level if different countries implement different measures. But additional policy measures are needed to reduce GHG emissions and internalise aviation's external costs. While we have pointed out several strategic arguments for airlines and airports to take action, we believe that policy measures are essential for actors to overcome an initial state of inertia. International regulative measures lower the fears of a prisoner's dilemma, help level the playing field, and serve management to justify investments deemed unnecessary by certain groups of shareholders. Policies will also bring about necessary behavioural change for customers, who have sometimes shown large discrepancies between their pro-environmental attitudes and their hypermobile behaviours.

First and foremost, policy measures need to be effective, but this can also include small measures. Examples of such measures include a declaration of emissions on tickets, defining how CO_2 charges have to be charged to the passenger (e.g. explicitly in addition to the final ticket price), regulating the marketing communication for airline offers (e.g. no price communication), and environmental education. Such measures serve to change the way our society thinks about mobility consumption, which is a required first step in diminishing demand. In addition, we call for regulators to foster the supply for SAF by ensuring demand for synthetic fuels, for example by implementing blending quotas (if availability is given) or subsidising production, as large-scale implementation of carbon-neutral fuels is essential to reduce GHG in the medium run.

In sum, we point to a wide range of measures policymakers have at hand. Unlike measures such as levies, taxes, and bans which typically bring strong opposition, the approaches suggested above can have a significantly positive effect on the climate impact of aviation and be more easily accepted. It is, however, imperative that policymakers know the interdependencies in the industry and understand the (unintended) negative impact that some policies may have. We, therefore, stress the need for coordinated measures that involve all relevant actors in the aviation system.

2.4 Airports

Like airlines, airports are also strongly influenced by policymakers and regulations. Airports have a responsibility to reduce their own scope 1 and 2 emissions. Here, airports have effective instruments at their disposal, which are already being successfully implemented. However, we argue that this influence is, firstly, limited in scope and, secondly, that the further potential for improvement of the airport infrastructure may soon be exhausted. Therefore, we see a much more important role for airports as central actors in aviation networks and in empowering airlines and customers to reduce their carbon footprint.

We identify 11 distinct levers airports have that empower other system members to achieve environmental sustainability. The framework is a simple but powerful tool that can be adapted to the airport's unique context. Examples of such measures include requirements for airlines to turn off engines during taxiing or to incentivise customers to use public transport to get to the airport—both measures which can reduce the scope 1 emissions for the respective actors.

The incentive for airports to do so is not purely altruistic but driven by tangible benefits such as extended utilisation rate of their assets, cost savings, reduced, or avoided damages from natural disasters as well as improved public reputation.

2.5 Airlines

After discussing measures in the other subsystems, the remaining subsystem of airlines has been touched upon several times already. Looking at the interdependencies and the influence all the actors have on aviation sustainability, airlines arguably face the biggest challenges. Trapped in a highly competitive market and dependent on strategic actions taken by peers, engaging in sustainability may seem like a burden for carriers. We take a different perspective and argue that keeping up with the changing environment and integrating sustainability into their core strategy is economically and strategically vital for airlines. Doing so should be seen as a significant opportunity to gain a competitive advantage.

Faced with a difficult post-COVID market and recovery, airlines are understandably in a place of inertia when it comes to sustainability that needs to be overcome.

Many airlines place the responsibility for climate actions elsewhere, for example with regulators or with overarching bodies such as CORSIA. Unfortunately, current strategic plans to mitigate air travel GHG emissions proposed by ICAO and IATA are not enough to achieve the Paris climate goals. Solely focusing on efficiency gains will fall short of these goals, since there is little more potential to gain in lowering emissions from contemporary technologies.

We put forward two general strategic options: consistency and sufficiency. Firstly, and to us the most promising strategy, is consistency. This means replacing current technologies with new sustainable ones. The main focus lies on synthetic aviation fuels. We acknowledge that there is (policy) support required to commercialise the technology, nevertheless to us it is the immediate way forward to drastically reduce air travel emissions without requiring many operational changes.

Secondly, sufficiency is an approach to further reduce GHG emissions. We believe that reducing and replacing point-to-point short-haul flights of up to 500 km with alternative, more sustainable modes of travel, such as high-speed rail, is necessary to work towards keeping global warming below 2 °C. While this does not directly decarbonise aviation, we see it as a likely political development (as recently implemented in France) that will affect airlines and further stress our claim that airlines seek strategic options in the broader context of a mobility provider. We are aware that a modal shift is not a viable alternative for most of the remaining long-haul flights. As a second-best option, even if the use of SAF will not definitely solve the climate problem there, it will at least improve it and we, therefore, argue for driving the development forward.

3 Outlook

What remains at the end of this book is a look into the future. In the different chapters, we tried to show that change at the global level is not only necessary but also urgent. Nevertheless, it is difficult to predict whether aviation will be able to keep its promises and achieve its (partly self-imposed) sustainability goals.

Primarily, the industry is still recovering from the COVID-19 shock. However, recent forecasts project an almost full recovery of the industry and indicate that recuperation in some regions of the world will be even faster than anticipated. IATA expects that, in retrospect, the pandemic will have been only a dampener on growth and that, for example Asia-Pacific and Africa/Middle East will return to growth rates (CAGR) above 5% between 2025 and 2030 (IATA, 2021). From a sustainability perspective, it is therefore essential to implement effective decarbonisation measures now, despite COVID, as otherwise the reduction targets will become increasingly difficult to achieve, as growth will outpace the reduction progress.

While we expect a continued growth in passenger numbers, at the same time, we are currently also seeing certain developments at a political level, which can serve as indicators for future attempts to slow down this growth of demand. Efforts to restrict aviation are gaining momentum in Europe. France, for example has decided to ban domestic flights if there are viable rail alternatives. The same call is on the table again in Germany. Recently, the Swiss electorate narrowly rejected a revision of the CO_2 law that would have included an air ticket levy. This shows that the discussion about politically imposed climate protection measures in aviation remains extremely current despite COVID-19. For the time being, these developments may mainly concern Europe. However, it is possible that eventually other regions of the world will also advocate stronger climate protection in aviation.

Even though predicting future developments is always a bit like reading in a crystal ball, we are convinced that there is an immediate need for action. A combination of technological progress and behavioural change, flanked by effective policy measures, will define the future development of aviation. Our recommendations for the action described above are by no means completely conclusive but should serve as a basis for discussion and inspiration for the players in the aviation system on how the industry can master the transition to a climate-neutral future.

We are convinced that the pressure to act will continue to increase and that the aviation industry will inevitably have to undergo profound change. From our point of view, industry players should be open also to fundamental changes and embrace them, as they always present strategic opportunities to prepare for the new, prospective market conditions in an early stage, instead of sticking to the status quo for too long. This will also include other modes of transport than flying.

In this sense, we hope that our book has helped to deepen the understanding of the topic "sustainable aviation" from a management perspective and that we have been able to present some possible solutions for the industry at the same time—coupled with an encouragement to see decarbonisation for aviation not only as a risk, but above all as an opportunity.

Reference

IATA Economics. (2021). *COVID 19 - An almost full recovery of air travel in prospect*. Retrieved from https://www.iata.org%2Fen%2Fiata-repository%2Fpublications%2Feconomic-reports%2Fan-almost-full-recovery-of-air-travel-in-prospect%2F&usg=AOvVaw0iQvfilc9RAiW4gkGufoEr

Closing Statement

Alexander Stauch and Adrian Müller

To conclude the book and its contents, we will add a few brief reflections and remarks on the process of the creation as well as on the outcome of the book. Writing this book was like an educational journey for us. At the beginning of the book project, there was only our own interest in bringing together the topics of travel and sustainability that are also both personally relevant to us. As rather young researchers with a strong passion for travelling, our motivation for the book project came from our fascination to travel on the one hand and on the other hand from the desire to find an intact environment and nature also in the future, which still allows travelling to far distant places. In other words: We wanted to find ways of still being able to enjoy travelling in the future without the environmental impact becoming so high that travelling itself will no longer be possible. A dilemma which is becoming increasingly relevant to many people our age, but also in other age groups.

With this mindset, our educational journey began. This journey included some highly relevant and new key-learnings to us, which will be briefly highlighted here. One of the most important findings was that technology alone will never be enough to make aviation climate neutral by 2050. In particular, too much focus on SAF will mean that the improvements achieved in the area of emissions will be overcompensated by the strong growth in demand. Moreover, as long as emissions are emitted whilst flying (which is the case with SAF and even hydrogen to some extent), there will always be an effect on global warming through radiative forcing. The real problem that stands in the way of sustainability in aviation is the expected strong increase in demand, which on the one hand will slow down an industry-wide transformation and on the other hand will make it very expensive as well. It is important for us to emphasise that we do not want to ban flying per se, but have built up a clear line of argumentation, which at the beginning also seemed new and drastic

A. Stauch
Institute for Economy and the Environment, University of St. Gallen, St. Gallen, Switzerland

A. Müller
University of St. Gallen, St. Gallen, Switzerland

to us, but which, according to all the findings, allows the only logical conclusion: we all have to fly less in the future (even if we all love to fly) and generally think more about our travel behaviour and the associated means of transport. We are aware that this finding is quite hard to take, especially if you are part of this industry and enthusiastic about travelling and flying. Thus, coming to this conclusion has taken quite some time until we were finally able to mentally accept it. But at the moment, it is the only way forward as new and disruptive technologies will not be available within the required timeframe. However, since not all people will think and reflect on this issue, we believe that demand management via supply prices is inevitable. This demand management can be carried out by the airlines themselves or initiated by politicians.

It is also important to emphasise that the time to act is now and must not be delayed any longer. The risks and dangers of irreversible climate damage for humankind and our beloved planet are too great to wait with the transformation now. Even if the sustainability goals of 2050 are still far in the future, they are still ambitious enough that we should start acting today rather than tomorrow.

By writing this book, we hope to make an important contribution to sustainable development and the transformation of aviation. We believe that this book can contribute decisively to new ways of re-thinking the industry and developing targeted measures that have a real impact in practice. Thus, we hope that the aviation industry can master the sustainable transformation and that we can still benefit from travelling to distant countries whilst keeping an intact environment also in the future.

Alexander Stauch is a Post Doc Researcher at the Institute for Economy and the Environment at the University of St. Gallen, who focused his research mainly on the area of consumer behaviour for renewable energies and e-mobility. Additionally, he also developed marketing concepts for new and sustainable products in the energy sector.

Adrian Müller is a PhD Student at the Center for Aviation Competence at the University of St. Gallen who focused his research mainly in the area of sustainable travel behaviour, with a particular interest in business travel. Additionally, he deals with different aviation topics from an academic perspective.

Printed by Printforce, United Kingdom